Decision
Theory

WILEY-INTERSCIENCE SERIES IN SYSTEMS AND OPTIMIZATION

Advisory Editors

Sheldon Ross

Department of Industrial Engineering and Operations Research, University of California, Berkeley, CA 94720, USA

Richard Weber

Statistical Laboratory, Cambridge University, 16 Mill Lane, Cambridge CB2 1RX, UK

The concept of a system as an entity in its own right has emerged with increasing force in the past few decades in, for example, the areas of electrical and control engineering, economics, ecology, urban structures, automation theory, operational research and industry. The more definite concept of a large-scale system is implicit in these applications, but is particularly evident in such fields as the study of communication networks, computer networks, and neural networks. The Wiley-Interscience Series in Systems and Optimization has been established to serve the needs of researchers in these rapidly developing fields. It is intended for works concerned with developments in quantitative systems theory, applications of such theory in areas of interest, or associated methodology.

Decision Theory

An Introduction to Dynamic Programming
and Sequential Decisions

John Bather

University of Sussex, UK

JOHN WILEY & SONS, LTD

Chichester · New York · Weinheim · Brisbane · Singapore · Toronto

Other Wiley Editorial Offices

New York • Germany • Brisbane • Singapore • Toronto

Library of Congress Cataloging-in-Publication Data

Bather, John
    Decision theory : an introduction to dynamic programming and sequential decisions
    / John Bather.
        p.   cm. — (Wiley-Interscience series in systems and optimization)
    Includes bibliographical references and index.
    ISBN 0-471-97649-0 (alk. paper)
     1. Decision Making 2. Dynamic Programming
     I. Title. II. Series.
T57.95. B34 2000
519.7'03—dc21   00-029620

British Library Cataloguing in Publication Data

A catalogue record for this book is available from the British Library

ISBN   0-471-97649-0

Produced from Postscript files supplied by the author
Printed and bound in Great Britain by Antony Rowe Ltd, Chippenham, Wiltshire
This book is printed on acid-free paper responsibly manufactured from sustainable forestry, in which at least two trees are planted for each one used for paper production.

To Sam and Emily,
Joshua and Ellie,
without whose entertaining distractions
I might have finished this book a good deal earlier.

Contents

CONTENTS

Preface

This book has evolved from a series of lecture courses I have given to mathematics students at the University of Sussex. The central aim of studying the optimization of sequences of decisions has remained the same, but there have been many changes over the years in order to introduce new topics and examples.

The material covered in Parts I and II is suitable for an optional course offered to final-year undergraduates in mathematics and statistics. There are few prerequisites: the calculus and real analysis usually covered in first-year courses and an introduction to probability should be enough. I hope the optimization problems investigated here will help to consolidate such basic ideas. Exercises are provided in Chapters 1 to 8 and notes on the solutions are given at the end of the book.

Although it is aimed primarily at mathematics undergraduates, I hope the book will also appeal to science and engineering graduates and to others working in the areas of optimization and operational research. Part III, consisting of Chapters 9, 10 and 11, is intended as an introduction to more advanced topics and the theory of Markov decision processes. Some of these topics are worth including at the end of an undergraduate course and Part III should also give some idea of the wide range of possible applications of the methods described earlier, without involving too much mathematical machinery.

Chapter 1 gives a discussion of the historical background and the relation between dynamic programming and mathematical induction. There is a brief summary of the contents of later chapters at the end.

Brighton
March, 2000

John Bather

1

Introduction

1.1 MATHEMATICAL INDUCTION

The principle of mathematical induction has been used for about 350 years. It was familiar to Fermat, in a disguised form, but the first clear statement seems to have been made by Pascal in proving results about the arrangement of numbers now known as Pascal's Triangle. This book contains many applications of inductive arguments and the aim here is to give some preliminary examples, illustrating why the method has become an indispensable tool in mathematics.

We begin with a general formulation of the principle. Let $p_1, p_2, p_3, \ldots$ be statements or propositions, each of which may be true or false.

The Principle of Induction

> *Suppose that*   (i) p_1 *is true*
> *and that, for* $n \geq 1$ (ii) $p_n \Rightarrow p_{n+1}$,
> *then* $p_1, p_2, p_3, \ldots$ *are all true.*

Perhaps the most familiar applications are concerned with proving statements like the following one.

Example 1.1

$$p_n : \qquad 1 + 2 + \cdots + n = \tfrac{1}{2}n(n+1).$$

Proof The statement p_1 means that $1 = \frac{1}{2}1(1+1)$, which is true. Now suppose that p_n is true for some $n \geq 1$. Then, by adding $(n+1)$ to each side of the equation p_n, we obtain

$$1 + 2 + \cdots + n + (n+1) = \tfrac{1}{2}n(n+1) + (n+1) = \tfrac{1}{2}(n+1)(n+2).$$

In other words, $p_n \Rightarrow p_{n+1}$. Then it follows from the principle that p_n is true for every $n \geq 1$. $\qquad\square$

The next example is slightly harder, but the argument is very similar.

Example 1.2

$$p_n : \qquad 1^2 + 2^2 + \cdots + n^2 = \frac{n(n+1)(2n+1)}{6}.$$

Proof In this case, p_1 reduces to the obvious statement that $1 = 1$. If p_n holds for some, $n \geq 1$, then we have

$$
\begin{aligned}
1^2 + 2^2 + \cdots + n^2 + (n+1)^2 &= \frac{n(n+1)(2n+1)}{6} + (n+1)^2 \\
&= \frac{(n+1)\{n(2n+1) + 6(n+1)\}}{6} \\
&= \frac{(n+1)(n+2)(2n+3)}{6}.
\end{aligned}
$$

The last expression is equivalent to the right hand side of the equation p_{n+1}, so we have shown that $p_n \Rightarrow p_{n+1}$.   □

1.2 HISTORICAL BACKGROUND

The historical material in this section is based on the book by Boyer [Boy68]. It is remarkable that Fermat published very little on the theory of numbers, but many of his penetrating ideas were noted in the margins of a 1621 edition of the *Arithmetica* of Diophantus. Some of his theorems were proved by an inductive method that he called *infinite descent* and he used it with great ingenuity. We can illustrate the method more simply by proving a well known classical result.

Example 1.3

$$\sqrt{2} \text{ is irrational.}$$

Proof We begin by assuming that

$$\sqrt{2} = \frac{k_1}{k_2},$$

where k_1 and k_2 are positive integers. This will lead to a contradiction which shows that $\sqrt{2}$ cannot be expressed as a ratio of integers.

Our assumption means that $k_1^2 = 2k_2^2$, so k_1 must be even and $k_1 > k_2$. Now write $k_1 = 2k_3$, so that $k_2^2 = 2k_3^2$ and we obtain

$$\sqrt{2} = \frac{k_2}{k_3}.$$

$$4k_3^2 = 2k_2^2$$
$$k_2^2 = 2k_3^2$$
$$\sqrt{2} = \frac{k_2}{k_3}$$

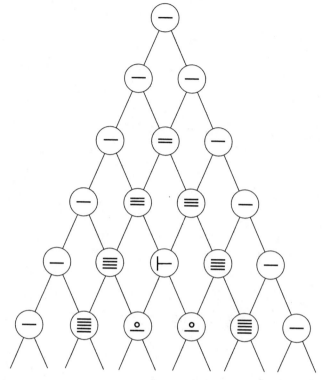

Figure 1. The Chinese triangle

By repeating this argument, we can obtain an equation

$$\sqrt{2} = \frac{k_n}{k_{n+1}},$$

for every $n = 1, 2, 3, ...$, where $k_1 > k_2 > k_3....$ This infinite descent is impossible, since the number of positive integers below k_1 is finite. □

Figures 1 and 2 give two different sketches of what is misleadingly called Pascal's Triangle. The first is a Chinese version from a diagram that appeared in the *Ssu-yüan yü-chien (Precious Mirror of the Four Elements)* by Chu Shi-chieh in 1303. Chu disclaims credit for the triangle and it seems likely that it originated in China about 1100. Note the use of rod numerals and the zero symbol in Figure 1. It is also interesting that formulae for the summation of series, like those in our first two examples, also appeared without proof in the *Precious Mirror*. Of course, both figures represent the same mathematical object. The reason that the triangle is associated with Pascal is that, in 1654, he gave a clear explanation of the method of induction and used it to prove some new results about the triangle. In fact, the construction of this infinite

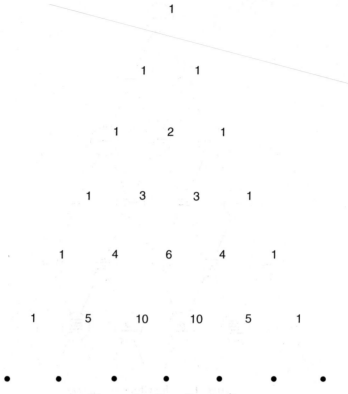

Figure 2.   Pascal's triangle

triangle is recursive so, with hindsight, inductive proofs now seem very natural.

Let the rth number in the nth row of Figure 2 be denoted by $\binom{n}{r}$ for $r = 0, 1, ..., n$ and $n = 0, 1, ...$. The first and last numbers in each row are both 1, so $\binom{n}{0} = \binom{n}{n} = 1$. The triangle is constructed by using the relation

$$\binom{n+1}{r+1} = \binom{n}{r} + \binom{n}{r+1} \tag{1.1}$$

for $r = 1, 2, ..., n-1$, to obtain the $(n+1)$th row from the nth. In other words, each entry is obtained by adding together the pair of numbers immediately above it in the previous row. It is a straightforward matter to establish the usual formula expressing $\binom{n}{r}$ in terms of factorials.

Example 1.4

$$\binom{n}{r} = \frac{n!}{r!\,(n-r)!} \tag{1.2}$$

Proof Let us take p_n to be the statement that equation (1.2) holds for $r = 0, 1, ..., n$. Note that p_0 and p_1 are both true, provided that 0! is interpreted as 1. Now suppose that p_n is true for some $n \geq 1$ and use the relation (1.1) to obtain

$$
\begin{aligned}
\binom{n+1}{r+1} &= \frac{n!}{r!\,(n-r)!} + \frac{n!}{(r+1)!\,(n-r-1)!} \\
&= \frac{n!}{(r+1)!\,(n-r)!} \{r+1+n-r\} \\
&= \frac{(n+1)!}{(r+1)!\,(n+1-(r+1))!} .
\end{aligned}
$$

The calculation is valid for $r = 0, 1, ..., n-1$ and we already know that $\binom{n+1}{0} = \binom{n+1}{n+1} = 1$. This completes the verification of p_{n+1}. $\qquad \square$

A final historical point is that the Principle of Induction was included in 1889 as one of Peano's axioms for the natural numbers, thereby recognizing it as one of the foundations of arithmetic.

1.3 DYNAMIC PROGRAMMING

The term *dynamic programming* was coined by Bellman [Bel57] to describe the techniques which he brought together to study a class of optimization problems involving sequences of decisions. There have been many applications and further developments since that time. A central aim of this book is to describe the method at an elementary level and to illustrate the range of possible applications.

Our subject is sequential, or multi-stage decision problems, where the time variable is used to order the sequence. We shall begin with deterministic examples which do not involve any random quantities or unknown parameters. As we shall see later, random variation can be included by taking expectations. Statistical problems are usually more difficult to formulate and solve, but it is worth mentioning that some of the techniques of dynamic programming first emerged in the investigation of sequential decisions by Wald and others. A comprehensive account of these statistical developments can be found in the book by DeGroot [DeG70].

Bellman deserves the credit for giving a clear statement of the principles of dynamic programming and for demonstrating a wide range of applications, but it would be misleading to suggest that the method is new since it relies heavily on mathematical induction. It turns out that it is natural to treat a sequence of decisions by reversing the order and, for this reason, the analysis is called *backwards induction*.

A deterministic model which covers many applications and a statement of Bellman's *principle of optimality* will be given in Chapter 2. It will be helpful to consider first some simple examples.

Example 1.5 The entries in the matrix

$$\begin{pmatrix} 2 & 5 & 3 & 8 & 6 \\ 4 & 2 & 9 & 4 & 1 \\ 5 & 3 & 2 & 6 & 9 \\ 0 & 3 & 8 & 5 & 0 \end{pmatrix}$$

represent costs associated with the positions in the rectangle. It is required to find an optimal route from the top left-hand corner to the bottom right-hand corner which consists of steps, either to the right or downwards, at each stage. The cost of following any particular route is the sum of all the entries encountered on the way. For example, the cost of moving down the first column and along the bottom row is $2 + 4 + 5 + 0 + 3 + 8 + 5 + 0 = 27$. An optimal route is one for which the total cost is a minimum.

In order to find such a route, we construct another matrix in which each entry represents the minimum cost, for that position, of reaching the bottom right-hand corner. The complete minimum-cost matrix is

$$\begin{pmatrix} 24 & 23 & 25 \rightarrow & 22 & 16 \\ \downarrow & \downarrow & \downarrow & \downarrow & \downarrow \\ 22 \rightarrow & 18 & 22 & 14 \rightarrow & 10 \\ & \downarrow & \downarrow & & \downarrow \\ 21 \rightarrow & 16 \rightarrow & 13 \rightarrow & 11 & 9 \\ \downarrow & & & \downarrow & \downarrow \\ 16 \rightarrow & 16 \rightarrow & 13 \rightarrow & 5 \rightarrow & 0 \end{pmatrix}$$

and the construction proceeds by examining the columns in reverse order. Thus, for the position in row 1, column 5, there is only one admissible path, as indicated by the arrows, and the minimum cost is $6 + 1 + 9 + 0 = 16$. We can now deal with positions in column 4, starting at the bottom where the minimum cost is $5 + 0 = 5$. The next entry in row 3, column 4, is obtained by noting that it is preferable to make the initial move downwards, giving a total cost $6 + 5 = 11$, rather than moving to the right. It does not take long to work backwards through the matrix in this way; the minimum cost for any position can be found by comparing the two neighbouring entries on the right and below it. The last entry to be obtained is that in the top left-hand corner and this is based on the fact that the smallest total of $2 + 22 = 24$ is achieved by making the first move downwards. Notice that the optimal route through the matrix is determined by following the arrows and, finally, we can verify the minimum total cost by adding up the appropriate entries in the original matrix: $2 + 4 + 2 + 3 + 2 + 6 + 5 + 0 = 24$.

A similar problem is that of finding the shortest path between two vertices in a network. The theory required for constructing such paths and a related technique called critical path analysis will be considered in Chapter 3. In order to introduce the basic ideas, suppose we have a network with vertices labelled

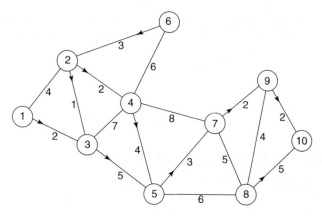

Figure 3. Network for Example 1.6

$1, 2, ..., n$, where the aim is to find the shortest path between vertex 1 and vertex n. Some, but not all, pairs of vertices are directly linked by an arc and the distance d_{ij} between i and j is given for these pairs. For simplicity, let us assume that there is at least one path between any two vertices: its length is obtained by adding the d_{ij} over the corresponding sequence of arcs. For any vertex i, the number of alternative paths from i to n is finite. Let f_i be the length of the shortest of these paths. Thus, $f_n = 0$ and it is not difficult to see that, for each $i < n$,

$$f_i = \min_j \{d_{ij} + f_j\}. \tag{1.3}$$

The minimization on the right of this equation is equivalent to choosing a direction of departure from vertex i: the index j runs over all vertices directly linked to i by an arc.

Suppose that all the given distances d_{ij} are positive. Then it can be shown that the system of equations (1.3) has a unique solution for the shortest lengths, $f_1, f_2, ..., f_{n-1}$, with $f_n = 0$. If we can determine this solution, then an optimal path from 1 to n can be found by following directions that attain the minimum in (1.3) at every vertex on the way. The solution is constructed by backwards induction: more precisely, the f_i are determined in increasing order. This is best demonstrated by looking at a particular case.

Example 1.6 Vertices $1, 2, ..., 10$ are shown in Figure 3 and each arc is marked with its length d_{ij}. The solution, giving the length of the shortest path to vertex 10 for each i, is found by working from right to left on the network:

$i =$	10	9	7	8	5	4	3	2	1	6
$f_i =$	0	2	4	5	7	11	12	13	14	16

The f_i satisfy equation (1.3) and, for each vertex i, the solution indicates a direction as shown by the arrows. Thus, the shortest path from 1 to 10 is

$$1 \rightarrow 3 \rightarrow 5 \rightarrow 7 \rightarrow 9 \rightarrow 10$$

and its length is given by

$$f_1 = 2 + 5 + 3 + 2 + 2 = 14.$$

1.4   THE EXECUTIONER'S TALE

We end this introduction with a cautionary tale. Inductive arguments are not always straightforward and the following anecdote contains one that is plausible, but false.

Many years ago, one Friday in court, a prisoner was convicted of a crime and sentenced to death. The executioner visited him in his cell and offered a hope of freedom. 'As it happens', he said, 'I am allowed some discretion in my work and I enjoy an occasional gamble. In your case, the execution is scheduled for next week and I have written the day: Monday, Tuesday, ... or Saturday, on a paper sealed in this envelope. I will visit you here early on Monday and then on the following days, if necessary, and ask whether you know the day of your execution. If you answer correctly at the first attempt, then you can go free, but otherwise I must do my job.'

The following Monday when the executioner arrived to ask his question, the prisoner replied immediately, 'Yes, it must be today.' 'What makes you say that?' said the executioner and this was the prisoner's argument. 'Consider the situation on Saturday morning. If you arrive then, I shall be certain that it is the appointed day, so it must be earlier in the week. Now consider Friday. If you ask me then, I will be sure of the answer because we have eliminated Saturday. Having excluded both the last two days, we can repeat the same argument for Thursday, and so on. By proceeding backwards in time we can eliminate the days until Monday is left as the only possibility.' The prisoner seemed quite satisfied with his conclusion and, to be fair, the executioner did not betray any emotion as he handed over the envelope to be opened – it was Wednesday!

There are several confusing features in the above argument; more than enough to invalidate the conclusion.

1.5   SUMMARY

Part I of this book is concerned with deterministic dynamic programming and the basic theory is described in Chapter 2. The mathematical model constructed there is in discrete time and it involves motion in a state space where the changes of state are controlled by a sequence of actions or decisions. Many sequential decision problems can be specified in this way, but the general model is introduced mainly to clarify the underlying ideas. There are many other problems that do not fit conveniently into this structure, but they can be investigated by using the same principles. In particular, the language and notation used to describe networks and critical path analysis in Chapter 3 are quite different, but the main ideas are similar. The following chapter also gives applications of discrete dynamic programming, each of which requires a slightly different approach. The method of backwards induction is developed further in Chapter 5 by showing how the special properties of convex and concave functions can be used to simplify the analysis of sequential decisions.

Some of the most interesting applications of dynamic programming involve models which include random variation. If all the random variables have known distributions, the model is called stochastic but, if there are unknown distributions, it is statistical. Part II of the book, consisting of Chapters 6, 7 and 8, extends the principles of dynamic programming to stochastic models. It is a straightforward matter to extend the model of Chapter 2 by introducing random variables and expectations. The aim is to minimize the total expected cost over the period of interest, since we cannot predict the exact costs associated with different decisions. Similarly, in other cases, it is appropriate to maximize expected rewards. The general principles are not seriously affected, but we now have a much wider range of possible applications. In Part II, we shall restrict attention to simple forms of decision procedure. Chapter 7 is concerned with optimal stopping problems, where the essential choice at each stage is whether to stop the underlying random process or allow it to continue. The special problems discussed in Chapter 8 also involve optimal stopping in various different settings. For example, in the marriage problem, sometimes called the secretary problem, the aim is to find an optimal policy for selecting the best from a sequence of candidates arranged in random order.

Part III is an introduction to more advanced topics. The theory of Markov decision processes was developed to deal with sequential decisions in a general model for random processes with discrete time and space variables. Chapters 9 and 10 describe methods for constructing policies that are optimal in the long run. This is done first by discounting future costs so that the total expectation over an infinite period of time is finite. However, discounting is not always appropriate and, in Chapter 10, another approach is described which leads to the minimization of the average cost over an infinite future. A full account of the theory and applications of Markov decision processes is contained in the recent book by Puterman [Put94].

The final chapter describes a statistical decision problem. Wald's fundamental work in sequential analysis [Wal47] led to the emergence of ideas very similar to the dynamic programming techniques being developed independently by Bellman. A key result in sequential hypothesis testing is the optimality of Wald's sequential probability ratio test and the proof obtained by Arrow, Blackwell and Girshick [ABG49] illustrates this similarity of ideas. The book by DeGroot [DeG70] covers the statistical background and gives a more detailed description of this and many other decision problems in statistics.

A comprehensive treatment of dynamic programming and its applications is contained in the two volumes by Whittle [Whi82], [Whi83]. They present a wide range of sequential decision problems in deterministic and stochastic control, including Markov decision processes and some statistical problems.

EXERCISES

1.1 Show that, for any positive integer n,

$$\frac{1}{2} + \frac{1}{6} + \frac{1}{12} + \cdots + \frac{1}{n(n+1)} = \frac{n}{n+1}.$$

1.2 Prove that

$$\binom{n}{0} + \binom{n}{1} + \binom{n}{2} + \cdots + \binom{n}{n} = 2^n.$$

1.3 Use the method of infinite descent to prove that $\sqrt{3}$ is irrational.

1.4 It is required to start at the $(1,1)$ position in the matrix below and proceed to the opposite corner, moving one step to the right or one step down at each stage, in such a way as to minimize the sum of the elements encountered. Find the optimal paths for this matrix.

$$\begin{pmatrix} 1 & 0 & 4 & -5 & 2 \\ 3 & 2 & 1 & 4 & -3 \\ 4 & 3 & -2 & 7 & 5 \\ 6 & -4 & -1 & 2 & 0 \end{pmatrix}$$

1.5 Sketch the network having 10 vertices and 15 arcs with lengths d_{ij} specified as follows:

$$\begin{array}{lllll}
d_{12} = 3, & d_{25} = 4, & d_{46} = 6, & d_{67} = 4, & d_{7,10} = 14, \\
d_{13} = 4, & d_{36} = 3, & d_{47} = 8, & d_{69} = 10, & d_{8,10} = 6, \\
d_{14} = 5, & d_{45} = 3, & d_{57} = 5, & d_{78} = 7, & d_{9,10} = 8,
\end{array}$$

Find the shortest path from vertex 1 to vertex 10.

1.6 Invent your own network and construct the shortest paths on it.

Part I

Deterministic Models

2

Multi-Stage Decision Problems

2.1 MAXIMIZING UTILITIES

We gave some simple examples of dynamic programming in Chapter 1 and the aim here is a more systematic development of the method. Let us begin with a case that is typical of many applications. It is concerned with maximizing utilities and it depends on a prescribed utility function U. This is supposed to represent the satisfaction obtained by a particular individual from spending money. When he spends an amount a, the corresponding utility is $U(a)$ and we shall investigate how to maximize the sum of all the utilities obtained over a number of spending decisions.

Suppose that $U(0) = 0$ and that the utility function has a derivative U' with

$$U'(a) > 0 \qquad (a > 0).$$

Thus, $U(a)$ is positive and strictly increasing in a.

Consider someone with initial capital x_0 and let $x_1, x_2, \ldots$ be the levels of capital at the times $1, 2, \cdots$. At time t, suppose he decides to spend an amount a_t within the range

$$0 \le a_t \le x_t. \tag{2.1}$$

The rest of his money is invested until time $t + 1$. More precisely, we assume that

$$x_{t+1} = \lambda(x_t - a_t), \tag{2.2}$$

where $\lambda > 1$ is a constant determined by the rate of interest. For example, if the unit of time is one year and interest is at 5%, then $\lambda = 1.05$. Over a period of length T, the total utility produced is

$$U(a_0) + U(a_1) + \cdots + U(a_{T-1}) \tag{2.3}$$

and the aim is to maximize this quantity, subject to the constraints imposed by relations (2.1) and (2.2). We are maximizing a function of T variables,

$a_0, a_1, \ldots, a_{T-1}$, which must be chosen under rather complicated conditions determined by the fact that x_t must remain non-negative for $t = 1, 2, \ldots, T$. Notice that any capital x_T, left over at the end of the period makes no contribution because no further spending is considered.

It is useful to define a new time variable

$$n = T - t, \tag{2.4}$$

which represents the time to go. This is also the number of decisions that remain to be taken before the period of interest ends at time T. The method of dynamic programming proceeds by induction on n. Consider the situation at time $t = T - 1$ when x_{t-1} becomes known. In this case $n = 1$. It is clear that our maximization problem is relatively simple when n is small.

In the first place, we shall focus on the special case of linear utilities. We begin by reformulating the maximization problem in terms of the time to go n and the initial level of capital x. It is important to realize that the maximum total utility obtainable from a sequence of spending decisions depends only on the number of decisions n and the initial capital x. This means that for any positive integer n and any real $x \geq 0$,

$$f_n(x) = \text{maximum total utility}$$

is properly defined. For a linear utility function, it is easy to guess the optimal policy, but it will be instructive to work through the backwards induction in detail.

Example 2.1

$$U(a) = a \qquad (a \geq 0). \tag{2.5}$$

Let us calculate $f_1(x)$ for an arbitrary x. Since there is only a single amount a to be chosen in the case $n = 1$, we have

$$f_1(x) = \max_{0 \leq a \leq x} a.$$

As might be expected, the amount spent should be as large as possible: $a = x$ and

$$f_1(x) = x.$$

In order to record the dependence of the optimal choice of a on n and x, we introduce a decision function d_n and write

$$a = d_n(x).$$

In particular, we have just worked out that

$$d_1(x) = x.$$

The case $n = 2$ is more typical of the inductive process. We observe that

$$f_2(x) = a + f_1(y),$$

where a now represents the amount spent in the first of two stages, x is the capital available at the start of the first stage and y is the capital for the second stage. Thus, a must be chosen within the range $0 \le a \le x$ and the choice determines

$$y = \lambda(x - a)$$

by using equation (2.2). Since we already know the maximum utility $f_1(y) = y$ in the final stage, we can evaluate

$$f_2(x) = \max_{0 \le a \le x} \{a + \lambda(x - a)\}.$$

The expression on the right here is

$$a + \lambda(x - a) = \lambda x - (\lambda - 1)a$$

and, since $\lambda > 1$, the coefficient of a is negative. Hence, the maximum is obtained by choosing it as small as possible: $a = 0$. Our results for the case $n = 2$ can be summarized as follows:

$$f_2(x) = \lambda x, \qquad d_2(x) = 0,$$

for all $x \ge 0$.

A similar argument can be used to evaluate $f_3(x)$, now that we know the function f_2 and so on. Let us assume that f_{n-1} is a known function, for some $n \ge 2$. Then, by considering the capital x available at the start of a period of length n and the amount a spent in the first stage, we obtain

$$f_n(x) = \max_{0 \le a \le x} \{a + f_{n-1}(\lambda(x - a))\}. \tag{2.6}$$

In fact, we can use this relation to show that

$$f_n(x) = \lambda^{n-1} x \tag{2.7}$$

for all $n \ge 1$ and $x \ge 0$.

It is easy to prove the general formula (2.7) by induction.

Proof Our previous calculations show that the formula is correct when $n = 1$ and $n = 2$. We now assume that, for some $n \ge 2$ and any amount $y \ge 0$,

$$f_{n-1}(y) = \lambda^{n-2} y.$$

On substituting this expression into equation (2.6), with $y = \lambda (x - a)$, we obtain

$$f_n (x) = \max_{0 \leq a \leq x} \left\{ a + \lambda^{n-1} (x - a) \right\}.$$

Since $\lambda^{n-1} > 1$, the coefficient of a in the expression on the right is negative and the maximum is given by setting $a = 0$. This leads immediately to the required formula (2.7). $\qquad\square$

Notice that, in view of the choice $a = 0$ in the above argument, the decision function corresponding to (2.7) is simply

$$d_n (x) = 0. \qquad (2.8)$$

This holds for every $n \geq 2$ but, as we remarked earlier

$$d_1 (x) = x.$$

The optimal policy is implicit in these decision functions and it only remains to interpret the results. We need to examine the sequence of values $x_0, x_1, \ldots, x_T$ determined by (2.2) when the optimal amounts $a_0, a_1, \ldots, a_{T-1}$ are substituted. Here, it is natural to work forwards in time so we use (2.4). Initially, $t = 0$, $n = T$ and the capital is x_0. Assuming that $T \geq 2$, (2.8) shows that $a_0 = 0$ and then we can apply (2.2), obtaining $x_1 = \lambda x_0$. Similarly, we determine a_1 in terms of x_1 and then find x_2. In fact, all the quantities $a_0, a_1, \ldots, a_{T-2}$ must be set equal to zero and we obtain, in succession,

$$x_2 = \lambda^2 x_0, \qquad x_3 = \lambda^3 x_0, \ldots, \qquad x_{T-1} = \lambda^{T-1} x_0.$$

Finally, the choice of a_{T-1} is determined by the decision function

$$d_1 : a_{T-1} = d_1 (x_{T-1}) = x_{T-1},$$

which means that $x_T = 0$. For the linear utility function, the optimal policy is to spend nothing until the final stage and then spend all the accumulated capital. The total utility obtained is equal to the final amount spent: $a_{T-1} = \lambda^{T-1} x_0$. This agrees with our general formula (2.7) which shows that the maximum total utility over the whole sequence of decisions is

$$f_T (x_0) = \lambda^{T-1} x_0.$$

Of course, the results we have obtained depend on the simple form of utility assumed in (2.5). As we shall see later, more interesting policies are obtained when the utility function is non-linear. We are now in a position to develop a much more general model for dynamic programming, but the approach and some of the notation will be the same.

2.2 A GENERAL MODEL

Imagine a dynamic system which moves through a sequence of states $x_0, x_1, x_2, \ldots$ at the times $0, 1, 2, \ldots$. We suppose that the motion is controlled by choosing a sequence of actions $a_0, a_1, a_2, \ldots$. There is a cost associated with each transition from one state to the next, and we are required to find a policy which minimizes the sum of these costs. A wide range of optimization problems can be specified in this way.

The state x_t and the corresponding action a_t at time t could be vectors of given dimensions and the method is sometimes applied to systems with more general state and action spaces. However, in what follows, both x_t and a_t will be real numbers. The essential features of the model are that, given x_t at time t, the choice of a_t determines both the next state x_{t+1} and the cost c_t of the transition from x_t to x_{t+1}. We suppose that two functions K and L are prescribed and that, in general,

$$c_t = K\left(x_t, a_t, t\right), \tag{2.9}$$

$$x_{t+1} = L\left(x_t, a_t, t\right). \tag{2.10}$$

Thus, K determines the *immediate cost* of any particular action, whereas L defines the *law of motion*. The problem is to choose the sequence of actions in order to minimize the total cost over a given period. More precisely, we shall restrict attention to a period of length T and use a criterion based on the sum

$$c_0 + c_1, + \cdots + c_{T-1}. \tag{2.11}$$

The aim is to minimize this quantity, for a given initial state x_0, by choosing $a_0, a_1, \ldots, a_{T-1}$. Since we are neglecting costs incurred after time T, it is clear that the actions $a_T, a_{T+1}, \ldots$ are irrelevant.

A key idea underlying the method of dynamic programming is the recognition that, at any time t during the period of control, we need only concern ourselves with minimizing the total future cost:

$$c_t + c_{t+1}, + \cdots + c_{T-1}.$$

The choice of present and future actions cannot possibly affect the past. This reduced problem involves the actions $a_t, a_{t+1}, \ldots, a_{T-1}$, so the effective number of decision variables is

$$n = T - t.$$

The minimization problem is simpler if n is small and the idea of separating past and future costs reveals that it is natural to work backwards in time, considering the cases $n = 1, 2, 3, \ldots$ in that order. Roughly speaking, we are able to consider the actions one at a time, but to do this we shall need to rely on the following principle.

The Principle of Optimality

An optimal policy has the property that, whatever the initial state and initial decision are, the remaining decisions must constitute an optimal policy with regard to the state resulting from the first decision.

Bellman introduced this principle in his book [Bel57], with only a brief comment by way of justification. It is true that 'a proof by contradiction is immediate', but some further explanation is needed. Suppose that the remaining decisions do not constitute an optimal policy as the principle claims. Then, starting at x_0, the policy determines a_0 and hence x_1, but the total future cost

$$c_1 + c_2, + \cdots + c_{T-1}$$

is not minimized. In other words, by changing the remaining decisions, this future cost can be reduced. However, any such changes can be included in the policy $a_0, a_1, a_2, \ldots a_{T-1}$ for the whole period, which shows that the original policy cannot be optimal. This is the required contradiction.

Another point worth noting is that an optimal policy may not exist: it may be that the minimum total cost cannot be attained exactly. This is why we refer to a principle, rather than a theorem. The essential idea is still valid even when it is necessary to deal with infima, rather than exact minima. Later, the principle will be used in a wide range of applications not covered by the present deterministic model.

Here we need to deal with many different infima, one for each possible state x and index n, and it is useful to regard these quantities as the values of a sequence of functions known as *minimum future cost functions*.

Definition 2.1 *For any state x and positive integer $n = T - t$,*

$$f_n(x) = \inf_{a_t, a_{t+1}, \ldots, a_{T-1}} \{c_t + c_{t+1} + \cdots + c_{T-1}\},$$

where $x_t = x$, $x_{t+1} = L(x, a_t, T - n)$ and so on.

This notation shows that the relevant variables are n and x. It is an abbreviation for the result of a complicated minimization and it provides a convenient summary of the future possibilities at each stage. We treat x as the initial state and then apply the principle of optimality in choosing the initial action. Of course, we cannot explicitly calculate each of the functions $f_1, f_2, \ldots$ for the general model defined by (2.9) and (2.10), but we can see how backwards induction works.

At each stage in the argument, we consider a typical transition from x to $y = L(x, a, T - n)$ generated by choosing the action a. For simplicity, let us assume that there is an optimal action in every case and denote this by

$a = d_n(x)$. Consider the case $n = 1$, where $t = T - 1$. Then there is only a single cost term $c_{T-1} = K(x, a, T - 1)$ and we have

$$f_1(x) = \inf_a K(x, a, T - 1).$$

This determines the function f_1 and we can proceed to the case $n = 2$. Here, we are concerned with two actions and, by using the principle of optimality, we may assume that the second of these is optimal when we consider choosing the first. Thus, for any initial state x and action a, the total cost is

$$K(x, a, T - 2) + f_1(y),$$

where $y = L(x, a, T - 2)$. Hence

$$f_2(x) = \inf_a \{K(x, a, T - 2) + f_1(L(x, a, T - 2))\}.$$

In general, we can determine f_n in terms of f_{n-1} by using the same argument:

$$f_n(x) = \inf_a \{K(x, a, T - n) + f_{n-1}(L(x, a, T - n))\}. \tag{2.12}$$

The optimal choice of a serves to define the corresponding *decision function:*

$$a = d_n(x).$$

However, it is important to realize that the method is based on the sequence of minimum future cost functions; the decision functions are a useful by-product in some, but not all, applications.

Sometimes, optimal actions do not exist and the decision functions are not properly determined. Strictly speaking, in such cases there is no optimal policy, but policies can be found to approximate the required infimum of the total cost. In other cases, the optimal policy may not be unique. However, these difficulties are not very important in practice. For example, the first cannot occur when the set of possible actions is finite; the infimum over a finite set is always a minimum. Many applications involve computations to evaluate minimum future cost functions, rather than explicit mathematical formulae. This means that states and actions must be restricted to a finite number of possibilities. The computations always determine a sequence of decision functions, at least in a discretized form.

[handwritten margin note: in practice, we always have optimal policy because inf over finite set is a minimum]

2.3   APPLICATIONS

We return to the problem of maximizing utilities described in Section 2.1, with a non-linear utility function. The model is specified by relations (2.1) and (2.2), where x_t is the level of capital at time t and a_t is the amount spent.

Example 2.2

$$U(a) = a^{\frac{1}{2}} \qquad (a \geq 0).$$ (2.13)

As before, the aim is to maximize the total utility.

The general model of Section 2.2 can be applied directly by defining the functions K and L as follows:

$$K(x_t, a_t, t) = -U(a_t),$$

$$L(x_t, a_t, t) = \lambda(x_t - a_t),$$

where $\lambda > 1$ is a constant. This determines the law of motion

$$x_{t+1} = \lambda(x_t - a_t)$$

exactly as before, and the total cost to be minimized is

$$-\{U(a_0) + U(a_1) + \cdots + U(a_{T-1})\}.$$

This is obviously equivalent to maximizing the total utility and we can avoid repetition of negative signs simply by redefining the functions $f_1, f_2, \ldots$ as maximum rewards, rather than minimum costs. Thus

$$f_1(x) = \max_{0 \leq a \leq x} U(a)$$

and for $n \geq 2$,

$$f_n(x) = \max_{0 \leq a \leq x} \{U(a) + f_{n-1}(\lambda(x-a))\}.$$ (2.14)

This is a special case of equation (2.12), apart from the change of sign introduced for convenience.

For Example 2.2, we have

$$f_1(x) = \max_{0 \leq a \leq x} a^{\frac{1}{2}} = x^{\frac{1}{2}}$$

and the corresponding decision function is

$$d_1(x) = x.$$

Then, by setting $n = 2$ in (2.14) and by using the formula we have obtained for f_1, we find that

$$f_2(x) = \max_{0 \leq a \leq x} \left\{a^{\frac{1}{2}} + \lambda^{\frac{1}{2}}(x-a)^{\frac{1}{2}}\right\}.$$

It is a simple exercise to maximize the expression on the right. The partial derivative with respect to a is

$$\tfrac{1}{2}a^{-\frac{1}{2}} - \tfrac{1}{2}\lambda^{\frac{1}{2}}(x-a)^{-\frac{1}{2}}.$$

This changes sign from positive to negative as a increases from 0 to x and it follows that the required maximum is determined by setting the derivative equal to zero. Hence

$$\lambda a = (x-a), \qquad a = x/(1+\lambda).$$

The decision function determined here is

$$d_2(x) = x/(1+\lambda).$$

When a is replaced by $x/(\lambda+1)$ in the above formula for f_2, the terms can be rearranged to give

$$f_2(x) = (1+\lambda)^{\frac{1}{2}} x^{\frac{1}{2}}.$$

It is a straightforward exercise to evaluate f_3 which has a similar form. In fact,

$$f_n(x) = \left(1 + \lambda + \lambda^2 + \cdots + \lambda^{n-1}\right)^{\frac{1}{2}} x^{\frac{1}{2}}, \tag{2.15}$$

$$d_n(x) = x/\left(1 + \lambda + \lambda^2 + \cdots + \lambda^{n-1}\right), \tag{2.16}$$

and both these formulae are valid for all $n \geq 1$ and $x \geq 0$. The results are not difficult to prove by induction from the recurrence relation (2.14). The argument is based on a calculation very similar to the one we have just carried out to evaluate f_2.

Finally, consider the optimal spending policy over T stages, starting with capital x_0. It follows from (2.16) with $n = T$ that

$$a_0 = d_T(x_0) = x_0/\left(1 + \lambda + \lambda^2 + \cdots + \lambda^{T-1}\right).$$

Then we find that $x_1 = \lambda(x_0 - a_0)$ and a short calculation shows that

$$a_1 = d_{T-1}(x_1) = \lambda^2 x_0/\left(1 + \lambda + \lambda^2 + \cdots + \lambda^{T-1}\right).$$

Thus, $a_1 = \lambda^2 a_0$ and it is not difficult to verify that $a_2 = \lambda^4 a_0$ and so on. This kind of spending pattern, in which the amounts increase by a factor of λ^2 at each stage, seems more sensible than the optimal policy we obtained in Example 2.1 for a linear utility function.

It is worth examining the behaviour of (2.15) and (2.16) when n becomes large. The assumption that $\lambda > 1$ implies that the geometric series $1 +$

$\lambda + \lambda^2 + \cdots$ diverges. Hence, $f_n(x)$ becomes infinite as $n \to \infty$ if $x > 0$. However, $d_n(x)$ converges to zero. At first sight, this suggests that a policy corresponding to the decision function

$$d(x) = 0$$

might be a good one, in the long run. But the implication is that no capital would ever be spent, so the total utility would be zero. Sometimes a more useful long-term policy can be found by letting $n \to \infty$.

The next application is a control problem which involves moving towards a target in discrete steps. It is a very simple example of linear control with quadratic costs. We shall need a slight extension of the model in Section 2.2 in order to include a terminal cost for missing the target. The terminal cost is specified by the function f_0 and the backwards induction starts with this.

Example 2.3

$$K(x_t, a_t, t) = ca_t^2, \quad L(x_t, a_t, t) = x_t + a_t, \quad f_0(x) = mx^2, \qquad (2.17)$$

where c and m are positive constants.

If we are allowed T steps from an initial position x_0, then we must choose $a_0, a_1, \ldots, a_{T-1}$ in order to minimize the sum of the control and terminal costs:

$$c\{a_0^2 + a_1^2 + \cdots + a_{T-1}^2\} + mx_T^2,$$

where $x_T = x_0 + a_0 + a_1 + \cdots + a_{T-1}$.

The recurrence relation (2.12) can be applied, with $n = 1$, in the special case given by (2.17).

$$f_1(x) = \min_a \left\{ ca^2 + m(x+a)^2 \right\}.$$

There is no restriction on the step length a here, so the minimum can be found by differentiation. This yields $a = -m(c+m)^{-1}x$, and hence

$$f_1(x) = cm(c+m)^{-1}x^2.$$

Notice that f_1 is a quadratic similar to f_0: we have replaced the constant m by $m_1 = cm(c+m)^{-1}$. When (2.12) is used again to determine f_2, the calculation is essentially the same:

$$f_2(x) = \min_a \left\{ ca^2 + m_1(x+a)^2 \right\}.$$

It follows immediately that the minimum occurs when $a = -m_1(c+m_1)^{-1}x$ and that

$$f_2(x) = m_2 x^2, \qquad m_2 = cm_1(c+m_1)^{-1}.$$

By repeating the argument, we find that,

$$f_n(x) = m_n x^2,$$ (2.18)

where

$$m_n = c m_{n-1} (c + m_{n-1})^{-1} \qquad (n \geq 1).$$

It is easily verified, by induction, that

$$m_n = cm (c + nm)^{-1}.$$ (2.19)

The decision function corresponding to (2.18) comes from the minimization used to evaluate f_n in terms of f_{n-1}:

$$a = -m_{n-1} (c + m_{n-1})^{-1} x.$$

This can be simplified, by using (2.19), to obtain

$$d_n(x) = -m (c + nm)^{-1} x.$$ (2.20)

Some of the details were omitted from the above calculations, but they are worth checking. Similarly, it is left as an exercise to work out the effects of applying the decision functions $d_T, d_{T-1}, \dots, d_1$ over a period of length T. It turns out that, for any initial position x_0, the optimal policy is to use steps of equal length:

$$a_t = -m (c + Tm)^{-1} x_0 \qquad (t = 0, 1, \dots, T-1).$$

Then the final position is

$$x_T = x_0 + \sum a_t = c (c + Tm)^{-1} x_0.$$

For our final application in this chapter, we return to maximizing utilities, as in Examples 2.1 and 2.2, but with a utility function that illustrates the complications which can easily arise in dynamic programming.

Example 2.4

$$U(a) = 1 - \frac{1}{1+a} \qquad (a \geq 0).$$ (2.21)

As before,

$$f_1(x) = \max_{0 \leq a \leq x} U(a).$$

Since $U'(a) > 0$, we must choose $a = x$ here and

$$f_1(x) = 1 - \frac{1}{1+x}.$$

Then, for two stages, the principle of optimality yields

$$f_2(x) = \max_{0 \le a \le x} \{U(a) + f_1(\lambda(x-a))\},$$

$$f_2(x) = \max_{0 \le a \le x} \left\{1 - \frac{1}{1+a} + 1 - \frac{1}{1+\lambda(x-a)}\right\}. \qquad (2.22)$$

Let $\frac{\partial}{\partial a}\{\cdot\}$ denote the partial derivative with respect to a on the right. We have

$$\frac{\partial}{\partial a}\{\cdot\} = \frac{1}{(1+a)^2} - \frac{\lambda}{(1+\lambda(x-a))^2}. \qquad (2.23)$$

This derivative is strictly decreasing in a and it is negative for all $0 \le a \le x$ if it is negative for $a = 0$, in other words if $(1 + \lambda x)^2 < \lambda$ or, equivalently, if

$$0 < x < \lambda^{-\frac{1}{2}} - \lambda^{-1}. \qquad (2.24)$$

In this case, the maximum in (2.22) occurs when $a = 0$ and we have

$$f_2(x) = 1 - \frac{1}{1+\lambda x}. \qquad (2.25)$$

Now suppose that condition (2.24) does not hold and $x \ge \lambda^{-\frac{1}{2}} - \lambda^{-1}$. Then the partial derivative in (2.23) is non-negative when $a = 0$ and, when $a = x$, it is $(1+x)^{-2} - \lambda$ which is always negative since $\lambda > 1$. Hence, the maximum in (2.22) must occur when $\frac{\partial}{\partial a}\{\cdot\}$ is zero. This means that

$$a = \frac{\left\{x - \left(\lambda^{-\frac{1}{2}} - \lambda^{-1}\right)\right\}}{1 + \lambda^{-\frac{1}{2}}}$$

and a short calculation shows that the formula for f_2 is

$$f_2(x) = 2 - \frac{\left(1 + \lambda^{\frac{1}{2}}\right)^2}{1 + \lambda + \lambda x}. \qquad (2.26)$$

Thus f_2 is given by two different formulae, (2.25) and (2.26), according to whether condition (2.24) holds or not. It can be verified that f_2 is a continuous function in spite of the two different expressions. However, the function f_3 obtained from it by carrying out another maximization must be even more complicated. The only feasible way to evaluate the minimum cost functions $f_3, f_4, \ldots$ is by computation.

EXERCISES

2.1 Consider the problem of maximizing utilities described in Section 2.1, with the utility function

$$U(a) = a^2 \qquad (a \geq 0).$$

Show that, if $\lambda > 1$, no matter what the initial capital or the number of stages involved, the optimal strategy is to save up until the final stage and then spend it all.

2.2 Prove the general formulae (2.15) and (2.16) given in Example 2.2.

2.3 A woman has capital x at the start of a period of two years. In the first year she spends an amount a and invests the rest, leaving $\lambda(x-a)$ to be spent in the second year, where $\lambda > 1$ is a given constant. Her utility function for spending in either year is given by

$$U(a) = 1 - e^{-a} \qquad (a \geq 0).$$

Find the optimal choice of a in the first year, $0 \leq a \leq x$, for any $x > 0$.

2.4 Verify that, in Example 2.3, the optimal control policy is to use T steps of equal length.

2.5 Let x_t be the position of a particle on the real line at time $t = 0, 1, \ldots, T$. The particle may be moved during $(t, t+1)$ any distance a to the right at a cost $a^2(1+t)$. Finally at time T, there is a reward given by $2x_T$. Show that, for any x_0, the optimal first step is to move to $x_0 + 1$ and find the optimal final position.

2.6 Given a starting point x on the real line, it is required to find an optimal route to the origin under the following conditions. The cost of a step of length a in either direction is $1 + a^2$ and any number of steps may be used to reach the point 0. Suppose first that the number of steps n is fixed in advance and prove that the optimal route is made up of n equal steps. Then show that the total cost can be minimized by choosing n so that $n(n-1) < x^2 \leq n(n+1)$.

2.7 Modify the control problem of Example 2.3 so that

$$K(x_t, a_t, t) = c|a_t|, \qquad L(x_t, a_t, t) = x_t + a_t$$

and let $f_0(x^2) = mx^2$, as before. Show that, for a single step, the minimum cost is given by

$$f_1(x) = mx^2 \qquad\qquad (|x| \leq c/2m),$$

$$f_1(x) = c|x| - c^2/4m \qquad\qquad (|x| \geq c/2m).$$

2.8 By using the Principle of Optimality and the two formulae for f_1 in the previous exercise, verify that $f_2(x) = f_1(x)$ for all x. Deduce that, for $n \geq 2$ steps,

$$f_n(x) = f_1(x)$$

always holds. What is the optimal control policy?

3

Networks

3.1 SHORTEST PATHS

The problem of finding shortest paths on a graph or network was introduced, for illustration, in Chapter 1. Suppose again that we have a network with vertices $1, 2, \ldots, n$. Certain pairs of vertices are linked by an arc and, for such pairs, the distance d_{ij} between i and j is given. A path is a sequence of arcs and its length is the sum of the corresponding d_{ij}. It is required to find the shortest path from vertex 1 to vertex n.

Example 3.1 Figure 1 shows a network with 10 vertices and 18 arcs. For each vertex i, consider all the possible paths from i to the target 10 and define f_i as the length of the shortest path. Then $f_{10} = 0$ and we can determine the values f_i, for $i \leq 9$, by solving the equations

$$f_i = \min_j \{d_{ij} + f_j\}. \tag{3.1}$$

The solution is found by working backwards through the network. It is easily verified that the following values satisfy (3.1) at every vertex $i \leq 9$.

i	=	10	9	8	7	6	5	4	3	2	1
f_i	=	0	5	7	10	10	14	19	18	20	21

The choice of j needed to attain the minimum in (3.1) corresponds to finding the optimal direction to take on leaving vertex i, as shown in Figure 1. Hence, the shortest path from 1 to 10 is

$$1 \rightarrow 3 \rightarrow 5 \rightarrow 6 \rightarrow 10$$

and its length is

$$f_1 = 3 + 4 + 4 + 10 = 21.$$

In order to establish that the method is reliable, we need to check that solutions of the equations (3.1) always correspond to optimal paths. Suppose

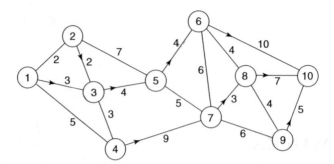

Figure 1.   Network for Example 3.1

that there are n vertices and that the specified arcs have lengths $d_{ij} > 0$. The aim, as before, is to reach vertex n. We assume that, for each $i < n$, there is at least one path:

$$i \to j_1 \to j_2 \to \cdots \to j_k = n.$$

Since the number of alternative paths from i to n is finite, this assumption guarantees the existence of a shortest path.

Proposition 3.1 *Consider the system of equations*

$$f_i = \min_j \{d_{ij} + f_j\},$$

where the minimization includes every vertex j which is linked to i by an arc. The equations have a unique solution for $f_i, i < n$, subject to the condition that $f_n = 0$, and f_i is the length of a shortest path from vertex i to vertex n.

Proof Let us define f_i to be the length of a shortest path from i to n and let $f_n = 0$. For $i < n$, consider the first arc $i \to j$ on any path from vertex i. The shortest possible path with this as its first arc has length $d_{ij} + f_j$, by the principle of optimality. Hence

$$f_i \leq d_{ij} + f_j \tag{3.2}$$

for every j which is directly accessible from i and equality must hold for some j, so we have

$$f_i = \min_j \{d_{ij} + f_j\}. \tag{3.3}$$

It remains to prove that this solution is unique. Let $g_i, i = 1, 2, \ldots, n$, be any other solution of (3.3) with $g_n = 0$. Given $i < n$, we can find a new vertex $j \neq i$ which attains the minimum in (3.3) for f_i:

$$f_i = d_{ij} + f_j.$$

Since the arc $i \to j$ need not be an optimal first choice for g_i, we have

$$g_i \leq d_{ij} + g_j,$$

and it follows that

$$g_i - f_i \leq g_j - f_j. \tag{3.4}$$

It is also clear that $f_i > f_j$, since $d_{ij} > 0$. The vertex i can now be replaced by j, which is closer to vertex n in the sense that $f_i > f_j$ and such that (3.4) holds. Then the argument can be repeated, starting with vertex j, and so on. This construction determines a path:

$$i \to j \to k \to \cdots \to n.$$

Vertex n must be attained in a finite number of steps, since $f_i > f_j > f_k > \cdots > f_n = 0$, and we also have

$$g_i - f_i \leq g_j - f_j \leq g_k - f_k \leq \cdots \leq g_n - f_n = 0.$$

We have shown that $g_i - f_i \leq 0$ for any $i < n$. Similarly, by interchanging the two solutions of (3.3), we can also prove that $f_i - g_i \leq 0$. Hence, $f_i = g_i$ always holds and there is only one solution of the equations. $\qquad\square$

3.2   DIRECTED NETWORKS

So far, we have been concerned with networks in which the links between vertices can be traversed in either direction. For certain problems, it is preferable to rely on one-way links called directed arcs. The use of directed networks has little effect on the problem of finding shortest paths, but it is important in critical path analysis, which is concerned with scheduling the various jobs in a construction project. As we shall see, this leads to the determination of longest paths through the network.

Consider a network with vertices $i, j = 1, 2, \ldots, n$ and a subset of the possible ordered pairs (i, j) which specifies the directed arcs. We write $i \to j$ if (i, j) is a directed arc. A path consists of a sequence:

$$i_1 \to i_2 \to i_3 \to \cdots \to i_k.$$

It will be assumed that the network is *acyclic*: in other words, there are no paths with $i_1 = i_k$. Then we can always rearrange the order of the vertices so that every directed arc $i \to j$ has $i < j$.

Proposition 3.2 *If a directed network is acyclic, then the vertices can be renumbered in such a way that $i < j$ for each directed arc $i \to j$.*

Proof There must be at least one vertex in the network which is not the end point of an arc. Otherwise, we can construct a cycle in the following way. Start at any vertex j and replace it by the initial point i of an arc $i \to j$. Then i is the end point of another arc, so we can replace it by the initial point and so on. This process generates a path by reversing the direction of each arc and it must eventually determine a cycle when one of the vertices is repeated.

Having identified a vertex which is not an end point, we assign the number 1 to it and then consider the reduced network obtained by removing this vertex and every arc of the form $1 \to j$. The reduced network is acyclic, so it must contain a vertex which is not an end point and this is renumbered as vertex 2. By successively reducing the number of vertices to $n-1, n-2, \ldots, 2, 1$ in this way we can rearrange the original network to obtain the required property.□

From now on, let us assume that we are dealing with an acyclic network and that every directed arc $i \to j$ has $i < j$. It follows that motion along any path in the network always increases the vertex number.

Suppose that a time $t_{ij} \geq 0$ is specified for each arc $i \to j$. In our previous interpretation of the model, this represents the time taken to move from i to j and we can minimize the total time required to reach vertex n from any initial vertex i by solving a system of equations similar to (3.3). Let $f_n = 0$ and define

$$f_i = \text{minimum time to reach } n$$

for $i < n$. The optimality equations are

$$f_i = \min_{j>i} \{t_{ij} + f_j\}. \tag{3.5}$$

Only those vertices j accessible by a directed arc $i \to j$ are included in the minimization here. This reformulation of the shortest path problem has the advantage that equations (3.5) can always be solved by backwards induction. Each f_i can be found as soon as all the values f_j for $j > i$ are known. Hence, we can use the equations to determine $f_{n-1}, f_{n-2}, \ldots, f_2, f_1$ in that order.

3.3 CRITICAL PATH ANALYSIS

The rest of this chapter is concerned with a different interpretation of directed networks. From now on, the arcs will represent activities or jobs in a construction project and the vertices will represent different stages reached during the completion of the project. Thus t_{ij} is the time interval required to complete the activity $i \to j$ and vertex j cannot be attained until at least t_{ij} after vertex i is reached.

Critical path analysis is a technique developed in operational research which has become a valuable tool in project management. It is used to schedule the jobs in large construction projects and in the preparation of bids or proposals

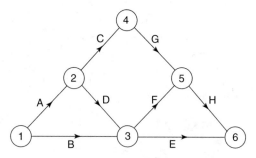

Figure 2.   Network for Example 3.2

for new projects. Books on operational research usually treat critical path analysis as an application of linear programming but we will continue to rely on a sequential approach.

A project is specified by a list of jobs, together with their precedence constraints, so a necessary preliminary is to convert these into a suitable network for analysis. Examples 3.2 and 3.3 below illustrate how this can be done. This and other practical matters are discussed in detail in most textbooks on operational research: see Ravindran, Phillips and Solberg [RPS87], for example. There is a brief sketch of the theory underlying critical path analysis in Whittle, Volume I [Whi82], and the aim in what follows will be to expand on the ideas mentioned there.

In a project network, the arcs represent jobs and their directions indicate the job sequences in the following way. It is assumed, for each vertex, that every job directed towards the vertex must be completed before any job directed away from that vertex can begin. Some examples will demonstrate how such networks are constructed.

Example 3.2 A project consists of eight jobs, $A, B, \ldots, H$, with the following constraints:

Job A precedes C and D

Jobs B and D precede E and F

Job C precedes G

Jobs F & G precede H

The network is shown in Figure 2. The project starts at vertex 1 and ends at vertex 6. The precedence constraints are represented by intermediate vertices. Sometimes it is necessary to introduce dummy jobs and vertices as illustrated in the next example.

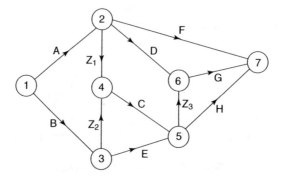

Figure 3.   Network for Example 3.3

Example 3.3 Again there are eight jobs and the constraints are:

> Job A precedes C, D and F
>
> Job B precedes C and E
>
> Jobs C and E precede G and H
>
> Job D precedes G

The network in Figure 3 contains dummy jobs Z_1, Z_2 and Z_3. These are assumed to have zero completion times and they are introduced so that constraints can be satisfied. For example, the dummy jobs Z_1 and Z_2 precede the real job C. In effect, vertex 4 marks the completion of both jobs A and B before C can begin.

Consider a project represented by a directed network which is acyclic and for which each vertex has at least one path leading from it to the final vertex. According to Proposition 3.2, we may assume that the vertices are numbered $1, 2, \ldots, n$ in such a way that all the directed arcs $i \rightarrow j$ have $i < j$. Let $t_{ij} \geq 0$ be the completion time required for the job represented by arc $i \rightarrow j$. For each arc $i \rightarrow j$ in the network, we shall say that i is a *predecessor* of j and that j is a *successor* of i. Let P_j be the set of predecessors of j and let S_j be the set of successors of j. Thus vertex j cannot be reached until at least a time t_{ij} after vertex i and this holds for every $i \in P_j$. Notice that P_j is a subset of the vertices $\{1, 2, \ldots, j-1\}$ and S_j is a subset of $\{j+1, j+2, \ldots, n\}$.

We now define g_j as the *early completion time* for vertex j and set $g_1 = 0$. Assuming that the project starts at time zero, for any $j > 1$, the earliest possible time that stage j can be attained is g_j. It is easy to see that an optimal policy for achieving these early completion times is to ensure that every job in the project starts as soon as all the jobs immediately preceding it are finished. This policy requires that, at any particular time, all the permissible jobs are carried on simultaneously without restriction on the necessary resources. The aim here is to complete the whole project in the shortest possible time g_n. As we shall see later, there are usually some jobs which can be delayed without affecting this final completion time. Critical path analysis is done in order to reveal which jobs are critical in the sense that they cannot be delayed.

The early completion times $g_2, g_3, \ldots, g_n$ can be found by an inductive method which moves forwards in time. It is clear that, for each j and every $i \in P_j$, we must have

$$g_j \geq t_{ij} + g_i.$$

If $g_1, g_2, \ldots, g_{j-1}$ are known, then we can determine the earliest time which satisfied these inequalities by finding the maximum of all the times on the right-hand side for $i \in P_j$. Hence,

$$g_j = \max_{i \in P_j} \{t_{ij} + g_i\} \tag{3.6}$$

holds for $j = 2, 3, \ldots, n$. We start with $g_1 = 0$; then $g_2 = t_{12} + 0$ and so on. Finally, we obtain g_n, the early completion time for the project.

It follows from our definition of the early completion times that they satisfy equation (3.6). To ensure that there are no other solutions of the equations, we need to impose a further condition. Vertex 1 has no predecessors, since it represents the start of the project, but there may be other vertices without predecessors. We can ensure uniqueness by setting $g_j = 0$ whenever the set P_j is empty.

Proposition 3.3 *The early completion times $g_1, g_2, \ldots, g_n$ are the unique solution of equations (3.6) subject to the condition that $g_j = 0$ whenever P_j is empty.*

Proof This is similar to the argument used to establish uniqueness in Proposition 3.1. Let $f_1, f_2, \ldots, f_n$ be another solution of (3.6) with $f_j = 0$ when P_j is empty. We shall prove that $f_j \leq g_j$ if P_j is non-empty. In this case, we can find a vertex $k \in P_j$ such that

$$f_j = t_{kj} + f_k = \max_{i \in P_j} \{t_{ij} + f_i\}.$$

Since the choice of k here is sub-optimal for g_j, we have

$$g_j \geq t_{kj} + g_k$$

and it follows that

$$g_j - f_j \geq g_k - f_k. \tag{3.7}$$

We also know that $k < j$, since P_j is a subset of $\{1, 2, \ldots, j-1\}$.

We have shown that either P_j is empty and $f_j = g_j = 0$, or there is another vertex $k < j$ such that (3.7) holds. This argument can be repeated to produce a finite sequence of vertices $j > k > \cdots > m$, say, where the last vertex has no predecessor and hence $f_m = g_m = 0$. It follows that

$$g_j - f_j \geq g_k - f_k \geq \cdots \geq g_m - f_m = 0.$$

In other words, $f_j \leq g_j$. This is valid for any j and, similarly, it can be shown that $f_j \geq g_j$. Hence, $f_j = g_j$ always and this confirms that the solution of equations (3.6) is unique.   □

In particular, the proposition shows that g_n, the early completion time for the project is uniquely determined. We are now in a position to introduce the *late completion times*. Let $h_n = g_n$ and, for each $j < n$, define the late completion time h_j as the latest possible time for stage j to be reached such that termination of the project can occur at the time given by g_n.

Consider any vertex $j < n$ and let $k \in S_j$. There is at least one path starting at j and ending at n, so S_j cannot be empty. The time interval between stages j and k must be at least t_{jk} to allow the completion of job $j \to k$. Hence $h_k - h_j \geq t_{jk}$ or, equivalently, $h_j \leq h_k - t_{jk}$. This holds for every $k \in S_j$, so the latest possible time at stage j which satisfies all these constraints is given by

$$h_j = \min_{k \in S_j} \{h_k - t_{jk}\}. \tag{3.8}$$

Since S_j is a subset of the vertices $\{j+1, j+2, \ldots, n\}$, this equation determines the value of h_j when the values h_k, for $j > k$, are known. It is clear that, given $h_n = g_n$, the late completion times $h_{n-1}, h_{n-2}, \ldots, h_1$ can be found by using equations (3.8). The method of Proposition 3.3 can be used to verify that there is a unique solution of these equations for $j < n$, subject to the condition that $h_n = g_n$.

We noted in the derivation of (3.8) that the late completion times satisfy inequalities of the form $h_k - h_j \geq t_{jk}$. For the same reasons, $g_k - g_j \geq t_{jk}$. Hence, for every arc $j \to k$ in the network, we have

$$g_k \geq g_j + t_{jk}, \qquad h_k \geq h_j + t_{jk}. \tag{3.9}$$

The same set of inequalities applies to both the early and the late completion times. In general, there are many other solutions but, as we might anticipate, the minimal solution of the inequalities (3.9) is provided by the early completion times, and the maximal solution is given by the late completion times: see Exercise 3.5.

A *critical path* can be defined as a path through the network from vertex 1 to vertex n such that

$$g_j - g_i = t_{ij}$$

for every arc $i \to j$ included in it. Consider any path

$$j_0 \to j_1 \to j_2 \to \cdots \to j_m$$

with $j_0 = 1$ and $j_m = n$. The inequalities (3.9) show that

$$g_{j_r} - g_{j_{r-1}} \geq t_{j_{r-1}, j_r} \qquad (3.10)$$

and by summing this along the path, we obtain

$$g_n - g_0 = \sum_{r=1}^{m} g_{j_r} - g_{j_{r-1}} \geq \sum_{r=1}^{m} t_{j_{r-1}, j_r} .$$

The sum on the right is the length of the path and $g_n - g_0 = g_n$ is the minimum completion time for the project. The length of a path through the networks cannot exceed g_n. However, if it is a critical path, equality holds in (3.10) for every arc on it and the summation shows that the path has length g_n. We have proved that a critical path is a longest path through the network.

At any vertex j, we have $h_j \geq g_j$. If $h_j > g_j$, we say that $h_j - h_i$ is the *slack* at vertex j. It means that a delay, not exceeding this amount, can be allowed at stage j of the project without affecting its final completion at the earliest possible time g_n. In practice, it is useful to locate vertices where slack occurs in order to provide flexibility in the allocation of resources to jobs and perhaps reduce their costs.

The evaluation of early and late completion times for a project network determines either a single time or a range of feasible times at each stage in the project, on the assumption that the total duration of the project must be as short as possible. The range of times for a particular job in the project is given by the amount of slack at the vertex where it starts. Its timing is fixed if there is no slack at this vertex. In particular, there is no flexibility at any of the vertices on a critical path.

Proposition 3.4 *There is no slack at any vertex on a critical path.*

Proof Let j be a vertex on a critical path and suppose, for contradiction, that $h_j > g_j$. Since $h_n = g_n$, by definition, we have $j < n$. Let $j = j_1 \to j_2 \to \cdots \to j_m = n$ be the section of the critical path leading from j to n. From (3.9), we obtain $h_{j_2} - h_{j_1} \geq t_{j_1, j_2}$ and we also have $g_{j_2} - g_{j_1} = t_{j_1, j_2}$, since the path is critical. It follows that

$$h_{j_2} \geq h_{j_1} + t_{j_1, j_2} = h_{j_1} - g_{j_1} + g_{j_2}.$$

But $h_{j_1} > g_{j_1}$, by assumption, so we obtain $h_{j_2} > g_{j_2}$. A repetition of the argument shows that $h_{j_3} > g_{j_3}$ and so on. This eventually leads to the inequality $h_{j_m} > g_{j_m}$, which contradicts the fact that $h_n = g_n$. This confirms that $h_j = g_j$ is the only possibility. □

It will be instructive to carry out some calculations of early and late completion times.

Example 3.4 Consider the project described in Example 3.2 and the corresponding network shown in Figure 2. Suppose that the completion times for the various jobs are as follows

Jobs	A	B	C	D	E	F	G	H
Times	3	7	4	5	8	5	3	4

The letters indicating jobs in Figure 2 can now be replaced by completion times : $t_{12} = 3$, $t_{13} = 7$, $t_{23} = 5$, and so on. We find the early completion times by solving the equations (3.6), starting with $g_1 = 0$. Then

$$
\begin{aligned}
g_2 &= 3 + g_1 = 3, \\
g_3 &= \max\{7 + g_1, 5 + g_2\} = 8, \\
g_4 &= 4 + g_2 = 7, \\
g_5 &= \max\{5 + g_3, 3 + g_4\} = 13, \\
g_6 &= \max\{8 + g_3, 4 + g_5\} = 17.
\end{aligned}
$$

The early completion time for the project is $g_6 = 17$ hours and we are now in a position to determine the late completion times by using (3.8), with $h_6 = 17$. We obtain

$$
\begin{aligned}
h_5 &= h_6 - 4 = 13, \\
h_4 &= h_5 - 3 = 10, \\
h_3 &= \min\{h_5 - 5, h_6 - 8\} = 8, \\
h_2 &= \min\{h_3 - 5, h_4 - 4\} = 3, \\
h_1 &= \min\{h_2 - 3, h_3 - 7\} = 0.
\end{aligned}
$$

We can compare the early and late completion times at each vertex.

j	=	1	2	3	4	5	6
g_j	=	0	3	8	7	13	17
h_j	=	0	3	8	10	13	17

The only vertex where there is slack is vertex 4 and the critical path is

$$1 \to 2 \to 3 \to 5 \to 6.$$

Its length is $3 + 5 + 5 + 4 = 17$, which is the shortest completion time for the project.

Example 3.5 Figure 4 describes a project network with 8 vertices and 13 jobs. The job completion times are shown on the corresponding directed arcs. The solution for the early and late completion times is given below. As before, the g_j are determined by (3.6) and, once $g_8 = 15 = h_8$ is known, the values of $h_j, j = 7, 6, \ldots, 1$, can be verified by using the equations (3.8). There are many details to be checked, even for such a simple network. For example,

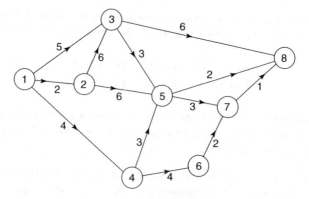

Figure 4. Network for Example 3.5

$$g_5 = \max\{6 + g_2, 3 + g_3, 3 + g_4\} = 11,$$
$$h_5 = \min\{h_7 - 3, h_8 - 2\} = 11.$$

Solution

j	=	1	2	3	4	5	6	7	8
g_j	=	0	2	8	4	11	8	14	15
h_j	=	0	2	8	8	11	12	14	15

There is some slack at vertices 4 and 6, and the critical path is

$$1 \to 2 \to 3 \to 5 \to 7 \to 8.$$

EXERCISES

3.1 Sketch the network having 9 vertices and 15 arcs, where the lengths are specified as follows:

$$d_{12} = 6, \quad d_{13} = 4, \quad d_{14} = 7, \quad d_{23} = 3, \quad d_{26} = 8,$$
$$d_{34} = 2, \quad d_{35} = 5, \quad d_{45} = 1, \quad d_{56} = 5, \quad d_{57} = 4,$$
$$d_{67} = 4, \quad d_{68} = 2, \quad d_{69} = 6, \quad d_{79} = 7, \quad d_{89} = 3.$$

Find the shortest path from vertex 1 to vertex 9. Indicate how you could prove that your path achieves the shortest possible length.

3.2 Now suppose that the network in the previous question defines a construction project. The 15 jobs are represented by directed arcs $i \to j$, with $i < j$, and their completion times are given by corresponding lengths. Find the early and late completion times at every stage of the project and determine any critical paths.

3.3 Consider the project network described in Example 3.3 and Figure 3. Suppose that the completion times for the eight jobs are as follows.

Jobs	A	B	C	D	E	F	G	H
Times	5	2	9	1	4	12	5	8

Calculate the early and late completion times and find the critical path. Show that the shortest path from vertex 1 to vertex 7 is half the length of the longest path.

3.4 Prove that the equations (3.8) have a unique solution for $h_1, h_2, \ldots, h_{n-1}$, subject to the condition that $h_n = g_n$, where g_n is the shortest completion time for the project.

3.5 Let $g_1 = 0, g_2, g_3, \ldots, g_n$ be the early completion times determined by (3.6) for a project network and let $h_1, h_2, \ldots, h_n$ be the late completion times obtained by solving (3.8), with $h_n = g_n$. Let $f_1, f_2, \ldots, f_n$ satisfy the conditions $f_1 \geq 0, f_n \leq g_n$ and

$$f_k \geq f_j + t_{jk}$$

for every directed arc $j \to k$ in the network. Prove that

$$g_j \leq f_j \leq h_j$$

for all j.

3.6 Invent your own directed network with, say, 10 vertices and determine both the shortest and longest paths for it. Find the lengths of several other paths through the network, for comparison.

4

Further Applications

4.1 DISCRETE ACTIONS

In the previous chapter, we were concerned with finding optimal decisions at the vertices of a network. At each stage, we had to choose from a finite set of actions representing the possible directions for moving on from the present vertex. The three different applications considered here are again concerned with discrete actions. In two of these cases, the state and time variables are not given explicitly so the problems do not fit immediately into the general model described in Chapter 2. This means that more care is needed in obtaining a suitable formulation, before the principles of dynamic programming can be used.

4.2 THE KNAPSACK PROBLEM

This is also known as the cargo loading problem. Suppose there are objects of several different types to be loaded into a container with limited capacity. The aim is to maximize the total value of the cargo, subject to a restriction on the total weight or volume. We assume that there are k types of item and that each item of type i has a specified positive value v_i and weight $w_i, i = 1, 2, \ldots, k$. Now suppose that the total weight permitted must not exceed w. The problem is to find non-negative integers $r_1, r_2, \ldots, r_k$ in order to

$$\text{maximize} \quad \sum_{i=1}^{k} r_i v_i,$$

$$\text{subject to} \quad \sum_{i=1}^{k} r_i w_i \leq w.$$

The restriction to integer values of the decision variables r_i is important. Otherwise, the solution to the problem is fairly obvious: we could fill up the knapsack, or ship, entirely with objects of a single type. For example, suppose that the types are arranged in decreasing order of value per unit weight. In other words, let

$$\frac{v_1}{w_1} \geq \frac{v_2}{w_2} \geq \cdots \geq \frac{v_k}{w_k}.$$

Write, $v_1 = \mu w_1$, so that $v_i \leq \mu w_i$ for all $i \leq k$. Then any admissible choice of the integers $r_1, r_2, \ldots, r_k$ yields a total value $\sum r_i v_i \leq \mu \sum r_i w_i \leq \mu w$. If we are allowed to choose $r_1 = w/w_1$ and set $r_i = 0$ for $i \geq 2$, then the total value is $r_1 v_1 = \mu w$ and the upper limit is achieved. But this can only occur if the ratio w/w_1 happens to be an integer. Such cases are rare and they give little indication of how to approach the problem more generally.

Suitable state and time variables for dynamic programming are not obvious from the specification of the knapsack problem. We shall treat the available capacity as our state variable and the number of different types of object as the time to go. Let x represent the available capacity, $0 \leq x \leq w$, and suppose that the different types of item are labelled $1, 2, \ldots, k$. For any $n \leq k$, we define $f_n(x)$ to be the maximum total value of a cargo made up from items of types $1, 2, \ldots, n$.

$$f_n(x) = \max \sum_{i=1}^{n} r_i v_i, \tag{4.1}$$

where each $r_i \geq 0$ and $\sum r_i w_i \leq x$.

We have

$$f_1(x) = \max \{rv_1 : rw_1 \leq x\}.$$

Since $v_1 > 0$, this maximum occurs when r is the integer part of x/w_1, written $r = [x/w_1]$. Thus, f_1 is a step function, increasing by v_1 at each successive multiple of w_1. We can evaluate f_2 in terms of f_1:

$$f_2(x) = \max \{rv_2 + f_1(x - rw_2)\}.$$

Here, $x - rw_2$ is the remaining capacity when r is chosen, and the optimal choice $r = r_2(x)$ is found by considering $r = 0, 1, \ldots, [x/w_2]$. The aim is to evaluate $f_k(w)$ and find optimal choices of $r_1, r_2, \ldots, r_k$. In effect, we need to calculate $f_n(x)$, for $n = 1, 2, \ldots, k$ and $0 \leq x \leq w$, by using the dynamic programming equation

$$f_n(x) = \max \{rv_n + f_{n-1}(x - rw_n)\} \tag{4.2}$$

to obtain $r = r_n(x)$ in the range $0 \leq r \leq [x/w_n]$.

Applications of (4.2) involve trying out all possible values of r for many different values of n and x, so the calculations can be tedious. We begin with a simple case.

Example 4.1 Suppose that the maximum weight allowed in our container is five units and suppose there are three different types of item. The weights and values of individual items are as follows.

i	w_i	v_i
1	2	8
2	3	10
3	1	3

In this case, there are only five distinct ways of filling the container, so it is easy to find the optimal choice of r_1, r_2, r_3 by direct enumeration. The dynamic programming approach determines the required maximum $f_3(5)$ in a more coherent way by first evaluating $f_1(x)$ and then $f_2(x)$ for $x = 1, 2, \ldots, 5$. It is helpful to use a table with the rows setting out values of f_1, f_2 and f_3:

x	1	2	3	4	5
f_1	0	8	8	16	16
f_2	0	8	10	16	18
f_3	3	8	11	16	19

For example, $f_2(4) = \max\{10r + f_1(4 - 3r)\}$ and by considering $r = 0$ or 1, we find that $f_2(4) = f_1(4) = 16$ with $r = 0$. This corresponds to choosing $r_2 = 0, r_1 = 2$. Similarly, when we evaluate $f_3(5)$, all the relevant values of f_2 are available and

$$f_3(5) = \max\{3r + f_2(5 - r)\}.$$

The possible choices $r = 0, 1, 2, 3, 4, 5$ produce total values 18, 19, 17, 17, 15, 15, respectively, so $r = 1$ and $f_3(5) = 19$. The optimal solution has $r_3 = 1$ and, since $f_3(5) = 3 + f_2(4)$, we also find that $r_2 = 0$ and $r_1 = 2$. It is easy to check that $\sum r_i v_i = 19$.

Example 4.2 Suppose there are four different types with the weights and values of single items given below.

i	w_i	v_i
1	6	25
2	5	20
3	3	11
4	2	7

The values of $f_n(x)$ obtained by using equation (4.2) are given in the following table, for $n \leq 4$ and $x \leq 10$:

x	1	2	3	4	5	6	7	8	9	10
f_1	0	0	0	0	0	25	25	25	25	25
f_2	0	0	0	0	20	25	25	25	25	40
f_3	0	0	11	11	20	25	25	31	36	40
f_4	0	7	11	14	20	25	27	32	36	40

For a specified maximum weight of 8 units, we have $f_4(8) = 32 = 7 + f_3(6) = 7 + f_1(6)$ and this total value is achieved by setting $r_1 = r_4 = 1$, $r_2 = r_3 = 0$. Similarly, if the total weight is 10, we can achieve $f_4(10) = 40$ by selecting $r_2 = 2$, $r_1 = r_3 = r_4 = 0$.

4.3   A SIMPLE REPLACEMENT MODEL

Imagine a useful article, such as a machine, which is needed frequently. Over a period of time, its performance deteriorates and it becomes more costly to maintain, so eventually it must be replaced by a new one. The process of wear and gradual deterioration over time is complex and any realistic model should allow random changes in the state of the machine as it is used. However, a very simple deterministic model is enough to illustrate how we can approach the problem of finding an optimal replacement policy. The examples described here show how a policy of regular replacements emerges as the best type of strategy for minimizing the average cost of maintenance and replacement in the long run. Later on, in Chapter 9, we will return to this problem and investigate a more realistic model.

Example 4.3 Consider a system with state variable x_t, which represents the age of the machine in use at time t. The state changes according to the rule:

$$x_{t+1} = x_t + 1, \text{ unless the machine is replaced, in which case } x_{t+1} = 0.$$

We assume that the cost of operating the machine over the next unit of time is cx_t and the cost of a replacement is k, where c and k are positive constants. Thus, the cost of a normal transition $x_t \to x_t + 1$ is cx_t and a replacement is represented by setting $x_{t+1} = 0$ at cost k.

Let $f_n(x)$ be the smallest possible total of all operating and replacement costs over a period of length n, starting with a machine of age x. Then $f_1(x) = \min\{cx, k\}$ for $x = 0, 1, 2, \ldots$ and, for $n \geq 1$,

$$f_{n+1}(x) = \min\{cx + f_n(x+1), k + f_n(0)\}. \tag{4.3}$$

It is easy to apply this dynamic programming equation to construct a table showing values of $f_1, f_2, \ldots$ for any particular choice of the parameters c and k.

For example, let $c = 1$ and $k = 7$. Then we obtain the following results:

x	0	1	2	3	4	5	6	7	8
f_1	0	1	2	3	4	5	6	7	7
f_2	1	3	5	7	7	7	7	7	7
f_3	3	6	8	8	8	8	.	.	.
f_4	6	9	10	10	10	10	.	.	.
f_5	9	11	12	13	13	13	.	.	.
f_6	11	13	15	16	16	16	.	.	.
f_7	13	16	18	18	18	18	.	.	.

It does not take long to see that there is a pattern in these results. Notice that f_7 is similar to f_3, apart from the addition of a constant:

$$f_7(x) = f_3(x) + 10.$$

This relation holds for all x and there is a similar relation between f_8 and f_4, and so on. If we add a constant to the function f_n on the right of equation (4.3), it does not affect the relative magnitude of $cx + f_n(x + 1)$ and $k + f_n(0)$. Hence, the decision whether to continue using the same machine or replace it is unaffected and the function f_{n+1} simply increases by the same constant as f_n. This argument demonstrates that

$$f_{n+4}(x) = f_n(x) + 10 \qquad (4.4)$$

for each $n \geq 3$ and all $x \geq 0$. Every integer $n \geq 3$ can be expressed in the form $n = 3 + r + 4s$, where $r = 0, 1, 2$ or 3 and $s = 0, 1, 2, \ldots$. A repeated application of (4.4) shows that

$$f_n(x) = f_{3+r}(x) + 10s. \qquad (4.5)$$

It follows that the optimal decision in state x, when $n = 3 + r + 4s$, can be found by considering the same state x when $n = 3 + r$. An examination of the cases $n = 3, 4, 5, 6$ in the table reveals that the rule:

replace if $x \geq 3$

yields an optimal decision. By extension, the same rule is optimal for all $n \geq 3$ and any x. Equation (4.5) shows that the total cost increases by 10 whenever n increases by 4, so the average cost per unit time is 2.5. It can be verified that this is the minimum average cost in the long run and that the policy of always replacing the machine as soon as it reaches age three attains this minimum.

Example 4.4 In order to see the effect of changing the replacement cost, consider the model described in the previous example with $c = 1$ and $k = 10$. The functions $f_1, f_2, \ldots$ are obtained by using relation (4.3), as before. It is a straightforward exercise to evaluate f_1, f_2, f_3, f_4 and verify that f_5 has the values shown below.

x	0	1	2	3	4	5
f_5	10	14	15	16	16	16
f_6	14	16	18	19	20	20
f_7	16	19	21	23	24	24
f_8	19	22	25	26	26	26
f_9	22	26	28	29	29	29
f_{10}	26	29	31	32	32	32
f_{11}	29	32	34	35	36	36
f_{12}	32	35	37	39	39	39
f_{13}	35	38	41	42	42	42

In this case, it takes a little longer for the pattern to emerge. We find that

$$f_{13}(x) = f_8(x) + 16$$

and it follows, by the same argument as before, that

$$f_{n+5}(x) = f_n(x) + 16 \qquad (4.6)$$

holds for each $n \geq 8$ and all x. The values shown in the table also indicate optimal decisions, according to whether $f_{n+1}(x) = c + f_n(x+1)$ or $f_{n+1}(x) = k + f_n(0)$. Sometimes, the minimum in (4.3) is attained by either decision and the choice is immaterial. An examination of the cases $n = 8, 9, \ldots, 12$ in the table shows that the rule

replace the existing machine as soon as $x \geq 4$

always produces an optimal decision. We can deduce from (4.6) that this rule is optimal for every $n \geq 8$ and all $x \geq 0$, since the pattern repeats itself. The total cost increases by 16 when n increases by 5 and it is not difficult to establish, using (4.6) that the average cost over a period of length n converges, as $n \to \infty$:

$$f_n(x) / n \to 16/5 = 3.2$$

for any fixed state x. In other words, the minimum average cost in the long run is 3.2. It can be verified directly that the policy of regular replacements at age four attains the minimum.

4.4 SCHEDULING PROBLEMS

Imagine a list of tasks or jobs, each of which requires processing by several machines. The main aim of scheduling is to arrange the jobs in a timetable which is optimal, in some sense. We might be trying to minimize the total time needed before all the jobs have been processed by the appropriate machines. Scheduling problems occur whenever the same machines are used for different types of production in a factory. Complexity grows rapidly with the numbers of jobs and machines, especially when there are constraints. For example, consider a long list of lecture courses, each of which must be delivered during a specified number of periods each week in one of many possible lecture rooms. Here, the jobs are courses and the machines are rooms with certain capacities. Constraints are imposed on the timetable because student numbers must never exceed room sizes.

This subject is an important branch of operational research. An introduction to the mathematical ideas is provided in the book by French [Fre82]. Unfortunately, very few scheduling problems have clear mathematical solutions. We shall restrict our attention to two simple cases.

Suppose there are n jobs and m machines. Example 4.5 below considers the case when $m = 1$. Then it is easy to find a permutation of the jobs which minimizes the total waiting time. Even when $m = 2$, it is not obvious how to determine an ordering of jobs to minimize the time needed to complete the processing of every job on both machines. The solution is known as Johnson's algorithm, and its optimality can be established by a dynamic programming argument.

Example 4.5 Suppose there are n jobs and one machine. Let $a_i > 0$ be the processing time for job i on the machine. The time required to complete all the processing is $\sum a_i$ and this does not depend on the order of processing. If they are processed in the order $1, 2, \ldots, n$, the waiting time for job k up to and including processing is $a_1 + a_2 + \cdots + a_k$. Hence, the total waiting time is

$$W = a_1 + (a_1 + a_2) + \cdots + (a_1 + a_2 + \cdots + a_n) \qquad (4.7)$$
$$W = na_1 + (n-1)a_2 + (n-2)a_3 + \cdots + a_n.$$

This total achieves its minimum when the jobs are arranged in increasing order of processing times:

$$a_1 \leq a_2 \leq \cdots \leq a_n. \qquad (4.8)$$

Proof Clearly, there is an optimal arrangement, so let us assume that the jobs are relabelled so that (4.7) gives the smallest total waiting time. Now suppose that at least one of the inequalities (4.8) does not hold. This means that $a_k > a_{k+1}$, for some $k < n$. Consider the effect of switching the order in which the jobs k and $k+1$ are processed. The total waiting time becomes

$$W' = na_1 + \cdots + (n+1-k)a_{k+1} + (n-k)a_k + \cdots + a_n,$$

and the difference is

$$W' - W = (n+1-k)(a_{k+1} - a_k) + (n-k)(a_k - a_{k+1}).$$

Thus $W' - W = a_{k+1} - a_k < 0$ and this contradicts the assumption that W is a minimum.

$\square$

4.5 JOHNSON'S ALGORITHM

Now suppose there are n jobs and two machines, A and B. Each job must be processed first on A, then on B, and the processing times are given. Let a_i, b_i be the processing times for job $i, i = 1, 2, \ldots, n$. We can arrange the jobs in any order for processing on machine A and the same order is retained for machine B. Delays may occur when jobs leave A, if B is still occupied with a previous job. The problem is to find a permutation of the jobs which reduces these delays and allows completion of all the processing in the shortest possible time.

The solution, roughly speaking, is to arrange that earlier jobs have short processing times on A and later jobs have short processing times on B. Let $c_i = \min(a_i, b_i)$ for each job in the set $S_0 = \{1, 2, \dots, n\}$. The algorithm is based on the following rules:

(i) Find $\min_{i \in S_0} c_i$ and choose a job j which attains this minimum.

(ii) If, $c_j = a_j$ process job j first ; if $c_j = b_j$, take it last.

(iii) Replace S_0 by $S_1 = S_0 - \{j\}$ and repeat the procedure on the reduced set of jobs.

Each iteration fills another position in the order of processing until the sequence is complete. In order to prove that this sequence is optimal, we shall use a switching argument similar to one given in Whittle, Volume I, [Whi82].

Let S be a set of jobs to be processed on machines A and B and let $t \geq 0$ be the length of time needed for machine B to complete the processing of previous jobs. The state of the system just before machine A begins on a new job is represented by the pair (S, t). Let $i \in S$ be chosen as the next job to be processed. Then, after a time a_i, when job i is ready to move to B, t is reduced to $\max(t - a_i, 0)$ and machine B is committed for a further time b_i. Thus, S is replaced by $S - \{i\}$ and t becomes t_i, where

$$t_i = \max(t - a_i, 0) + b_i = \max(t - a_i + b_i, b_i).$$

For any set S and time $t \geq 0$, we define $M(S, t)$ to be the minimum time needed to complete the processing of all the jobs in S, given that machine B is already committed for a time t. The optimality equation is

$$M(S, t) = \min_{i \in S} \{a_i + M(S - \{i\}, t_i)\}. \tag{4.9}$$

We define $M(\emptyset, t) = t$ when the set S is empty. For any S, it is clear that $M(S, t)$ is a non-decreasing function of t.

Consider the effect on the state (S, t) of processing two jobs $i, j \in S$ in that order. The new state is $(S - \{i, j\}, t_{ij})$, where t_{ij} is given by

$$t_{ij} = \max(t_i - a_j + b_j, b_j).$$

By substituting for t_i here and rearranging terms, we obtain

$$t_{ij} = \max(t + b_i - a_i + b_j - a_j, \quad b_i + b_j - a_j, b_j).$$

On the other hand, if we process j first and then i, the new state is $(S - \{i, j\}, t_{ji})$ and the only change is that t_{ij} is replaced by

$$t_{ji} = \max(t + b_i - a_i + b_j - a_j, \quad b_i + b_j - a_i, b_i).$$

What matters here is whether $t_{ij} \leq t_{ji}$ or not. Note that the first of the three expressions on the right of t_{ij} coincides with the first expression in t_{ji}. Hence, $t_{ij} \leq t_{ji}$ if

$$\max(b_i + b_j - a_j, b_j) \leq \max(b_i + b_j - a_i, b_i).$$

If is a straightforward exercise to check that this inequality is equivalent to

$$\min(a_i, b_j) \leq \min(a_j, b_i). \tag{4.10}$$

We can establish the optimality of Johnson's algorithm by applying the following result.

Proposition 4.1 *Let S be any subset of $\{1, 2, \ldots, n\}$ with at least two members and let $t \geq 0$. Choose $i \in S$ such that $\min(a_i, b_j) \leq \min(a_j, b_i)$ holds for all $j \neq i, j \in S$. Then the minimum completion time $M(S, t)$ can be attained by a schedule starting with job i.*

Proof Consider an optimal schedule for the jobs in S and suppose that i is not the first job. Then there is another job $j \in S$ which precedes it in the optimal schedule and the inequality (4.10) holds. Let (S', t') be the state reached when all the jobs preceding j in the schedule have been processed by machine A. Since the rest of the schedule is optimal for the reduced set S', we must have

$$M(S', t') = a_j + a_i + M\left(S' - \{j, i\}, t'_{ji}\right),$$

where t'_{ji} is obtained from t' after machine A has processed job j and then i. The inequality (4.10) shows that $t'_{ij} \leq t'_{ji}$, and hence

$$M\left(S' - \{i, j\}, t'_{ij}\right) \leq M\left(S' - \{j, i\}, t'_{ji}\right).$$

The optimality of the schedule means that we cannot reduce the minimum completion time, so equality is the only possibility here. It follows that

$$M(S', t') = a_i + a_j + M\left(S' - \{i, j\}, t'_{ij}\right).$$

In other words, the schedule can be modified by interchanging jobs i and j without affecting its optimality. This switching argument can be repeated and eventually, it leads to an optimal schedule which starts with job i. □

This result can be used for the initial state $(S_0, 0)$, where $S_0 = \{1, 2, \ldots, n\}$, and for any subsequent state (S, t) which can arise. The implications of the proposition are that we can find an optimal schedule for S_0 by rearranging the jobs so that, for every pair i, j, job i precedes job j only if the inequality (4.10) holds. For convenience, suppose that the jobs are relabelled so that

$$\min(a_i, b_j) \leq \min(a_j, b_i), \quad \text{whenever } i < j.$$

Then the proposition shows that there is an optimal schedule starting with job 1. We can also apply it to the new state (S_1, t_1), where $S_1 = \{2, 3, \ldots, n\}$ and $t_1 = b_1$, which shows that it is optimal to process job 2 next, and so on.

We are now able to prove that Johnson's algorithm always leads to an optimal schedule. Suppose that the rules (i), (ii) and (iii) are applied repeatedly to rearrange the n jobs in a certain order. We must verify that (4.10) is satisfied whenever i precedes j in this schedule. If job i precedes job j, the rules mean that either $c_i \leq c_j$ and $c_i = a_i$, or $c_j \leq c_i$ and $c_j = b_j$. The two possibilities are that

$$\text{either} \quad a_i = \min(a_i, b_i) \leq \min(a_j, b_j),$$
$$\text{or} \quad b_j = \min(a_j, b_j) \leq \min(a_i, b_i).$$

It follows that (4.10) is satisfied. This establishes the optimality of the schedule.

Example 4.6 Suppose there are nine jobs with the processing times given below. This artificial case makes it easy to compare the best and the worst possible schedules. The ordering of the jobs is already optimal, according to Johnson's algorithm. For this schedule, machine B is committed for a time t_i when job i leaves machine A. We have $t_0 = 0$ and $t_i = \max(t_{i-1} + b_i - a_i, b_i)$, $i \geq 1$. The table also shows $c_i = \min(a_i, b_i)$.

a_i	1	2	3	4	5	6	7	8	9
b_i	9	8	7	6	5	4	3	2	1
c_i	1	2	3	4	5	4	3	2	1
t_i	9	15	19	21	21	19	15	9	1

The minimum completion time for all the processing on both machines is $\sum a_i + t_9 = 1 + 2 + \cdots + 9 + 1 = 46$.

For comparison, consider the effect of reversing the optimal schedule. The results are as follows:

a_i	9	8	7	6	5	4	3	2	1
b_i	1	2	3	4	5	6	7	8	9
c_i	1	2	3	4	5	4	3	2	1
t_i	1	2	3	4	5	7	11	17	25

In fact, this is the worst possible ordering of the jobs and we find that $t_9 = 25$, so the total completion time becomes $\sum a_i + t_9 = 70$.

EXERCISES

4.1 Solve the knapsack problem when the maximum weight allowed is five units and there are three types of item. The weights and values of individual items are as follows.

i	w_i	v_i
1	2	65
2	3	80
3	1	30

4.2 A manager has to decide how to distribute 10 salespeople between three marketing areas. The profit to be obtained from an area increases with the number of salespeople allocated to it according to the table below.

Number	0	1	2	3	4	5	6	7	8	9	10
Area 1	50	60	80	105	115	130	150	160	165	170	175
Area 2	50	65	85	110	140	160	175	185	190	195	200
Area 3	60	75	100	120	135	150	175	190	195	200	205

How many salespeople should be allocated to each area in order to maximize profits?

4.3 Consider the simple replacement model with $c = 1$ and $k = 4$. Show that $f_{n+3}(x) - f_n(x)$ does not depend on x, for $n \geq 3$, and deduce an optimal replacement policy which minimizes average costs in the long run.

4.4 Let $c = 1$ in the simple replacement model. For any fixed $r \geq 1$, let π_r be the policy of replacing the current machine as soon as its age reaches the critical level r. By considering the costs incurred over any period of length $r + 1$, show that the average cost per unit time for the policy π_r is

$$a_r = \frac{r(r-1) + 2k}{2(r+1)},$$

where $k > 0$ is the replacement cost. Prove that a_r can be minimized by choosing the largest integer r such that $r(r+1) \leq 2(k+1)$.

4.5 Ten jobs have the processing times a_i, b_i given below in random order. Find the time needed to complete all the processing on machines A and B when the jobs are taken in this order. Then apply Johnson's algorithm and obtain the minimum completion time.

a_i	13	11	18	20	20	19	11	20	19	3
b_i	3	5	8	19	6	8	15	10	9	5

4.6 Repeat the previous question when each job is processed on machine B first and then on machine A. Notice that in this case, Johnson's algorithm produces no improvement.

4.7 Let $a_1, b_1, a_2, b_2, \ldots, a_n, b_n$ and t be non-negative numbers and let

$$t_{ij} = \max\left(t + b_i - a_i + b_j - a_j, \ b_i + b_j - a_j, \ b_j\right)$$

for $i \neq j$. Show that the inequality (4.10) implies that $t_{ij} \leq t_{ji}$. Verify that if the jobs are rearranged by using Johnson's algorithm, then $t_{ij} \leq t_{ji}$ whenever i precedes j.

5

Convexity

5.1 CONVEX AND CONCAVE FUNCTIONS

This chapter is concerned with multi-stage decision problems covered by the general model introduced in Chapter 2. The dynamic programming approach involves a sequence of functions f_1, f_2, ... , where $f_n(x)$ represents the best that can be achieved over n stages when the initial state is x. The aim is to determine the functions f_1, f_2, ... , successively but, as we noted in Example 2.4, it may not be feasible to obtain explicit formulae. However, in such cases, it is sometimes possible to establish general properties of these functions which tell us something about the optimal decision rules and help to reduce the computations needed to find them.

Convexity is a property which can be useful in this way. This section gives the definitions of convex and concave functions and a brief description of their behaviour. A more comprehensive treatment can be found in books on convex analysis, such as the one by van Tiel [van84]. We shall return to dynamic programming applications in Sections 5.2 and 5.3.

Consider functions defined on the real line or on an interval contained in it.

Definition 5.1 *A real-valued function h is said to be convex if*

$$h(p_1 x_1 + p_2 x_2) \leq p_1 h(x_1) + p_2 h(x_2) \tag{5.1}$$

for any x_1, x_2 in its domain and any $p_1, p_2 \geq 0$ with $p_1 + p_2 = 1$.

The function h is said to be strictly convex if the strict inequality holds in (5.1) whenever $x_1 \neq x_2$ and both $p_1, p_2 > 0$.

Figure 1 illustrates what convexity means from a geometric point of view. The chord joining any two points $(x_1, h(x_1))$ and $(x_2, h(x_2))$ must always lie above the curve showing the graph of h.

We say that a function h is *concave* if (5.1) is replaced by the reverse inequality $\geq$, and h is *strictly concave* if $>$ holds always. Thus, h is concave if $-h$ is convex. This fact means that we can easily obtain results about concave functions from the corresponding properties of convex functions and we shall concentrate on the latter.

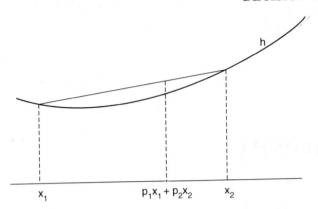

Figure 1. The chord property

The chord property used in the definition of convex functions has many implications and we need to derive several results for later applications.

Proposition 5.1 *Let h be a convex function defined on some interval I and let $x_j \in I, j = 1, 2, \ldots, n$. Then, for any $p_1, p_2, \ldots, p_n \geq 0$ with $\sum p_j = 1$,*

$$h\left(\sum_1^n p_j x_j\right) \leq \sum_1^n p_j h(x_j). \tag{5.2}$$

Proof This is true for $n = 2$, by definition of convexity. The inequality can be extended by induction. Assume that it is valid for any $n - 1$ points x_j and any $p'_j \geq 0$ such that $\sum_1^{n-1} p'_j = 1$. Now consider the point $x = p_1 x_1 + p_2 x_2 + \cdots + p_n x_n$ and suppose that $0 < p_n < 1$ and $n \geq 3$, which excludes only trivial cases. It is clear that $x \in I$ since it must lie between $\min(x_j)$ and $\max(x_j)$. We can also express it as a combination of two points in I:

$$x = (1 - p_n) x' + p_n x_n,$$

where

$$x' = p'_1 x_1 + p'_2 x_2 + \cdots + p'_{n-1} x_{n-1}, \qquad p'_j = p_j / (1 - p_n).$$

Note that $\sum p'_j = 1$ and so, by our inductive assumption

$$h(x') \leq \sum_1^{n-1} p'_j h(x_j).$$

From Definition 5.1, we have

$$h(x) = h((1 - p_n) x' + p_n x_n) \leq (1 - p_n) h(x') + p_n h(x_n).$$

Then, by using the preceding inequality,

$$h(x) \leq (1 - p_n) \sum_{1}^{n-1} p'_j h(x_j) + p_n h(x_n).$$

Since $(1 - p_n) p'_j = p_j$ for $j < n$, we have

$$h\left(\sum_{1}^{n} p_j x_j\right) \leq \sum_{1}^{n} p_j h(x_j),$$

which completes the proof.

$\square$

Proposition 5.1 is known as Jensen's inequality. It can be expressed in terms of expectations with respect to a random variable X, which takes the possible values $x_1, x_2, \ldots, x_n$ with probabilities $p_1, p_2, \ldots, p_n$, respectively. Thus, (5.2) is equivalent to

$$h(E\{X\}) \leq E\{h(X)\}. \tag{5.3}$$

In fact, Jensen's inequality holds for any random variable and any convex function. It can be shown that (5.3) remains valid whenever X is a random variable with finite expectation, taking values in the domain of h.

The sum of two convex functions is convex. The same holds for the maximum of two convex functions, but not for their minimum.

Proposition 5.2 *Let $h_1(x)$ and $h_2(x)$ be convex functions of x. Then*

(i) $h_1(x) + h_2(x)$,
(ii) $max\{h_1(x), h_2(x)\}$

are also convex functions.

Proof Both results follow easily from the Definition 5.1. Consider (ii), for example, and let h be the function defined by

$$h(x) = \max\{h_1(x), h_2(x)\}.$$

We need to verify that h satisfies the inequality (5.1) and this can be done by using the corresponding inequalities for h_1 and h_2:

$$
\begin{aligned}
h(p_1 x_1 + p_2 x_2) &= \max\{h_1(p_1 x_1 + p_2 x_2), h_2(p_1 x_2 + p_2 x_2)\} \\
&\leq \max\{p_1 h_1(x_1) + p_2 h_1(x_2), p_1 h_2(x_1) + p_2 h_2(x_2)\}.
\end{aligned}
$$

However, $h_i(x) \leq h(x)$ for $i = 1, 2$ and any x, so each of the two expressions on the right is bounded above by $p_1 h(x_1) + p_2 h(x_2)$. Hence

$$h(p_1 x_1 + p_2 x_2) \leq p_1 h(x_1) + p_2 h(x_2),$$

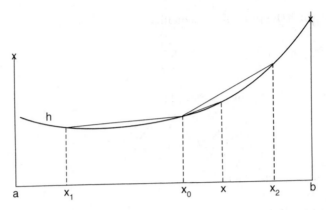

Figure 2.   Chords and one-sided derivatives

as required.   □

It can be shown, without difficulty, that the sum or the maximum of a finite number of convex functions is convex: see Exercise 5.1.

Proposition 5.3 *Let h be a convex function defined on a closed interval $[a,b]$ and let x_0 be an interior point : $a < x_0 < b$. Then h is continuous at x_0 and it has one-sided derivatives on the left and on the right of x_0, such that*

$$h'(x_0-) \le h'(x_0+).$$

Proof Choose fixed points such that $a < x_1 < x_0 < x_2 < b$ and let x be a variable point with $x_0 < x < x_2$: see Figure 2. By using the convexity of h, we can compare the slopes of chords joining different pairs of points on the graph of h. Thus,

$$\frac{h(x_0) - h(x_1)}{x_0 - x_1} \le \frac{h(x) - h(x_0)}{x - x_0} \le \frac{h(x_2) - h(x_0)}{x_2 - x_0}. \tag{5.4}$$

Now let $x \downarrow x_0$ from above. The inequalities show that the ratio $\{h(x) - h(x_0)\}/(x - x_0)$ is bounded by two quantities which remain fixed as x varies. It follows that $h(x) \to h(x_0)$ as $x \downarrow x_0$. A similar argument shows that $h(x) \to h(x_0)$ as $x \uparrow x_0$, which means that h is continuous at x_0.

Again consider the ratio $\{h(x) - h(x_0)\}/(x - x_0)$ as x decreases towards x_0. Another application of the chord property shows that this ratio changes monotonically. The inequality (5.4) gives a lower bound, so it must converge to a limit which defines the one-sided derivative.

$$\frac{h(x) - h(x_0)}{x - x_0} \downarrow h'(x_0+) \quad \text{as } x \downarrow x_0.$$

It is a straightforward matter to show further that the slope of a chord on the left of the point $(x_0, h(x_0))$ converges to a limit $h'(x_0-)$ and it is clear from comparisons of slopes that

$$h'(x_0-) \leq h'(x_0+),$$

as required.

□

It is worth noting that a function h which is convex on a closed interval need not be continuous at the end points, where upward jumps can occur: see Figure 2. However, if h is continuous at an end point, its one-sided derivative exists there. It is differentiable at an interior point x_0 if its one-sided derivatives are equal, but convexity does not guarantee this. For example, a convex function h attains its minimum value at x_0 if and only if

$$h'(x_0-) \leq 0 \leq h'(x_0+).$$

It is relatively easy to find the extreme values of convex or concave functions.

Proposition 5.4 *Let h be a convex function defined on the closed interval $[a, b]$ and suppose it is continuous at a and b. The following results hold:*

    (i)   *The maximum of h is attained at a or b:*
          $\max_{a \leq x \leq b} h(x) = \max \{h(a), h(b)\}.$

    (ii)   *The minimum occurs at a if $h'(a+) \geq 0$.*
          The minimum occurs at b if $h'(b-) \leq 0$.

    (iii) *If $h'(a+) < 0 < h'(b-)$, then the minimum of h*
          is attained at an interior point x_0, where
          $h'(x_0-) \leq 0 \leq h'(x_0+).$

Proof (i) This follows from the chord property, since it is contradicted if $h(x) > \max \{h(a), h(b)\}$ for some $x \in (a, b)$.
(ii) We know from the proof of Proposition 5.3. that the ratio $\{h(x) - h(a)\}/(x - a)$ is non-decreasing in x and it has a limit $h'(a+)$ as $x \downarrow a$. Hence,

$$h(x) - h(a) \geq (x - a) h'(a+),$$

and if $h'(a+) \geq 0$, we have $h(x) \geq h(a)$ for all $x \in [a, b]$. Similarly, we obtain

$$h(b) - h(x) \leq (b - x) h'(b-).$$

If, $h'(b-) \leq 0$, this shows that $h(x) \geq h(b)$ holds for all x.
(iii) Since h is continuous on the closed interval $[a,b]$, its minimum is attained at some point x_0. However, the inequalities $h'(a+) < 0 < h'(b-)$ show that it

cannot be minimized at either a or b. Hence, $a < x_0 < b$ and $h(x) - h(x_0) \geq 0$ always holds. In particular, this applies as $x \uparrow x_0$, or as $x \downarrow x_0$, and we must have $h'(x_0-) \leq 0 \leq h'(x_0+)$.   □

The above results (ii) and (iii) do not guarantee that the minimum is unique. For example, a convex function can remain constant at its minimum value throughout an interval. On the other hand, if h is strictly convex, its minimum in any closed interval occurs at a single point.

Roughly speaking, a convex function h has a derivative $h'(x)$ which is non-decreasing in x, apart from upward jumps at certain points. At worst, such jumps can only occur on a countable set of points, so $h'(x)$ exists almost everywhere. In general, we can rely on the inequalities satisfied by its one-sided derivatives:

$$h'(x_1-) \leq h'(x_1+) \leq h'(x_2-) \leq h'(x_2+),$$

whenever $x_1 < x_2$.

There is a simple test for convexity which can be applied to functions with second derivatives. It is easily verified that h is convex if $h''(x) \geq 0$ always holds.

Concave and strictly concave functions have properties very similar to those we have established in this section for convex functions. In the applications which follow, we shall make use of concave utility functions to represent the idea of diminishing returns.

5.2   ALLOCATION PROBLEMS

A merchant has a quantity A of a commodity and she intends to distribute it between N customers. The net return for sending an amount a_j to customer j is $R_j(a_j)$. Her problem is to choose the amounts $a_1, a_2, \ldots, a_n$ with each $a_j \geq 0$ and $\sum a_j = A$, such that the total return $\sum R_j(a_j)$ is maximized.

Time is not explicitly involved here, but the customers may be served in any order. For convenience of notation, let us imagine that they are served in the order $N, N-1, \ldots, 2, 1$. Then we can define $f_n(x)$ to be the maximum possible return from distributing an amount x between the last n customers. For any $n = 1, 2, \ldots, N$ and any $x \geq 0$

$$f_n(x) = \max \{R_n(a_n) + R_{n-1}(a_{n-1}) + \cdots + R_1(a_1)\}$$

subject to $a_j \geq 0$ and $a_n + a_{n-1} + \cdots + a_1 = x$. This sets up the problem for dynamic programming. We are mainly interested in $f_N(A)$, but we need to consider $f_n(x)$ for $n = 1, 2, \ldots, N$ and $0 \leq x \leq A$.

The standard technique described in Chapter 2 shows that the optimality equation is

$$f_n(x) = \max_{0 \leq a \leq x} \{R_n(a) + f_{n-1}(x-a)\}. \tag{5.5}$$

Clearly $f_1(x) = R_1(x)$ for all $x \geq 0$. In principle, we can evaluate the functions $f_2, f_3, \ldots$ by using (5.5). Note that an optimal choice of a gives the amount allocated to customer n: $a_n = d_n(x)$. As we shall see, if all the reward functions R_j are concave or if they are all convex, it is not difficult to find an optimal allocation policy.

Example 5.1 Suppose that all the reward functions are the same:

$$R_j(a) = R(a),$$

where R is a strictly concave function. For simplicity, suppose further that $R(a)$ is continuous at $a = 0$ and that it has a continuous derivative $R'(a)$ for $a > 0$. Then strict concavity implies that $R'(a)$ is strictly decreasing in a. For $n = 2$, we have

$$f_2(x) = \max_{0 \leq a \leq x} \{R(a) + R(x - a)\},$$

since $f_1 = R$. The expression on the right has a partial derivative with respect to a given by $R'(a) - R'(x - a)$ and this is strictly decreasing in a. It has a unique zero where $a = x - a$, so the maximum total return is obtained when $a = x/2$. Hence

$$f_2(x) = 2R\left(\frac{x}{2}\right), \qquad d_2(x) = \frac{x}{2}.$$

Consider the inductive hypothesis that

$$f_n(x) = nR\left(\frac{x}{n}\right) \tag{5.6}$$

holds for some $n \geq 1$ and all $x \geq 0$. By using the optimality equation (5.5) with n replaced by $n + 1$, we obtain

$$f_{n+1}(x) = \max_{0 \leq a \leq x} \left\{R(a) + nR\left(\frac{x - a}{n}\right)\right\}.$$

In this case also, we can locate the maximum by differentiation with respect to a. The derivative is

$$R'(a) - R'\left(\frac{x - a}{n}\right),$$

and it has a unique zero yielding the maximum when $a = (x - a)/n$, i.e. $a = x/(n + 1)$. It follows that

$$f_{n+1}(x) = (n + 1)R\left(\frac{x}{n + 1}\right).$$

This completes the proof of (5.6), for all $n \geq 1$ and $x \geq 0$.

The above proof also shows that $f_{n+1}(x)$ is attained by the allocation $d_{n+1}(x) = x/(n+1)$ to customer $n+1$. Thus x is replaced by $nx/(n+1)$ as the amount left for the remaining n customers. It is easily verified that the optimal allocation for the customer n is also $x/(n+1)$. In fact, the allocations are all equal. In particular,

$$f_N(A) = NR\left(\frac{A}{N}\right),$$

and the maximum total return is achieved by allocating an amount A/N to each customer.

The same allocation rule is optimal, but perhaps not unique, for any concave reward function R. The assumption that R has a continuous derivative is not necessary, although more care is needed when $R'(a)$ might have downward jumps at certain points. The formula for $f_N(A)$ can be interpreted in a different way, by noting that any admissible allocation must satisfy $\sum a_j = A$. We have shown that, for a concave function R and any amounts $a_j \geq 0$,

$$R(a_1) + R(a_2) + \cdots + R(a_N) \leq NR\left(\frac{a_1 + a_2 + \cdots + a_N}{N}\right).$$

In other words,

$$\frac{1}{N}\sum R(a_j) \leq R\left(\frac{\sum a_j}{N}\right). \tag{5.7}$$

This is, of course, a special case of Jensen's inequality (5.3): let $h(x) = -R(x)$ for $x \geq 0$ and define X to be the random variable which takes the values $a_1, a_2, \ldots, a_N$ with equal probabilities.

Example 5.2 Now suppose that each reward function $R_j(a)$ is convex in $a \geq 0$ and let $R_j(0) = 0$, which allows for possible discontinuities at $a = 0$. As we shall see, the optimal allocation rule is quite different: the merchant should send the whole amount A to a single customer.

We have $f_1(x) = R_1(x)$ and then (5.5) gives

$$f_2(x) = \max_{0 \leq a \leq x} \{R_2(a) + R_1(x - a)\}.$$

Both $R_2(a)$ and $R_1(x - a)$ are convex functions of a on the interval $0 \leq a \leq x$. By Proposition 5.2, so is their sum and Proposition 5.4 shows that the maximum is attained at $a = 0$, or at $a = x$. When $a = 0$, we obtain $R_2(a) + R_1(x - a) = R_1(x)$, since $R_2(0) = 0$. Similarly, $a = x$ yields a total return $R_2(x)$. Thus,

$$f_2(x) = \max\{R_1(x), R_2(x)\},$$

for all $x \geq 0$.

Note that $f_2(x)$ is a convex function, according to Proposition 5.2. It is helpful to include convexity in the inductive hypothesis used to establish the general formula for $f_n(x)$.

Suppose that, for some $n \geq 2$,

$$f_n(x) = \max\{R_1(x), R_2(x), \ldots, R_n(x)\} \tag{5.8}$$

and that this is a convex function for $x \geq 0$.

In order to verify that f_{n+1} has similar properties, we use the optimality equation (5.5).

$$f_{n+1}(x) = \max_{0 \leq a \leq x}\{R_{n+1}(a) + f_n(x - a)\}.$$

Since f_n is convex, by the inductive hypothesis, $f_n(x - a)$ is convex in a for fixed x. Hence we have a sum of two convex functions and, as in the case $n = 1$, the maximum is attained at either $a = 0$ or $a = x$. In other words,

$$f_{n+1}(x) = \max\{f_n(x), R_{n+1}(x)\}.$$

Finally, we can substitute for $f_n(x)$, using (5.8). The resulting formula is equivalent to

$$f_{n+1}(x) = \max\{R_1(x), R_2(x), \ldots, R_{n+1}(x)\}.$$

The previous equation shows that $f_{n+1}(x)$ is the maximum of two convex functions, so it is also convex. This completes the induction.

We have established that, for convex reward functions,

$$f_N(A) = \max\{R_1(A), R_2(A), \ldots, R_N(A)\}.$$

The merchant can maximize her total net return by transporting the amount A to one customer and nothing to the others. The choice of the best customer usually depends on A. This is illustrated by the following case.

Example 5.3 Let $R_j(0) = 0$ and, for $a > 0$, let

$$R_j(a) = p_j a - c_j,$$

where p_j and c_j are positive constants. Here p_j represents the price paid by customer j and c_j is the delivery cost incurred by the merchant, which does not depend on a. The most profitable single customer must be found by comparing the net returns $p_j A - c_j$, for $j = 1, 2, \ldots, N$. For example, in the special case $p_j = 2j$ and $c_j = j^2$, it can be shown that the net return is maximized by choosing customer k such that k is the integer nearest to the amount A, from the set $\{1, 2, \ldots, N\}$: see Exercise 5.4.

5.3   CONCAVE UTILITY FUNCTIONS

We return to the simple model of Chapter 2. The state variable $x \geq 0$ represents the level of capital. A decision to spend an amount a, $0 \leq a \leq x$, produces an immediate utility $U(a)$ and the next state y reached after investing the amount $x - a$ for one unit of time is

$$y = \lambda(x - a),$$

where $\lambda > 1$ is a constant determined by the rate of interest. In the dynamic programming formulation: see relation (2.14), $f_n(x)$ is the maximum total utility attainable over n units of time starting in state x. The optimality equation is

$$f_n(x) = \max_{0 \leq a \leq x} \{U(a) + f_{n-1}(\lambda(x - a))\}, \tag{5.9}$$

$n \geq 1$, with $f_0(x) = 0$, $x \geq 0$.

In this section, we shall concentrate on utility functions $U(a)$ which are increasing in a and concave. We noted in Example 2.4 that it may not be possible to obtain explicit formulae for $f_1(x), f_2(x)$, etc. However, it is sometimes possible to establish general properties of these functions. If the utility function is concave, then so are the functions, f_n, for all $n \geq 1$.

Example 5.4 Let $U(a)$ be a non-decreasing, concave function of $a \geq 0$ which is continuous at $a = 0$, with $U(0) = 0$. Then the functions $f_n(x)$ defined for $n \geq 1$ by equation (5.9) have similar properties. In particular, each $f_n(x)$ is concave and non-decreasing in $x \geq 0$.

Proof It is clear that $f_n(0) = 0$ always holds and it is easy to check that $f_n(x) \to 0$ as $x \downarrow 0$, by using a suitable upper bound on the possible amounts $a_1, a_2, \ldots, a_n$. We have $f_1(x) = U(x)$ for $x \geq 0$, from (5.9).

Suppose that $f_{n-1}(y)$ is concave and non-decreasing in $y \geq 0$. We need to establish that $f_n(x)$ also has these properties. Let $x_2 > x_1 \geq 0$ and consider

$$f_n(x_2) = \max \{U(a) + f_{n-1}(\lambda(x_2 - x_1) + \lambda(x_1 - a))\}.$$

The maximum here, over the range $0 \leq a \leq x_2$, cannot increase if we restrict to $0 \leq a \leq x_1$. The assumed properties of f_{n-1} guarantee that

$$f_{n-1}(\lambda(x_2 - x_1) + \lambda(x_1 - a)) \geq f_{n-1}(\lambda(x_1 - a))$$

holds for any choice of a. Hence,

$$f_n(x_2) \geq \max_{0 \leq a \leq x_1} \{U(a) + f_{n-1}(\lambda(x_1 - a))\} = f_n(x_1).$$

It remains to prove that f_n is a concave function. Let $x = p_1 x_1 + p_2 x_2$, where $x_1, x_2 \geq 0, p_1, p_2 > 0$ and $p_1 + p_2 = 1$. We must show that

$$f_n(x) \geq p_1 f_n(x_1) + p_2 f_n(x_2). \tag{5.10}$$

Suppose that $f_n(x_i) = U(a_i) + f_{n-1}(\lambda(x_i - a_i))$ for $i = 1, 2$, where a_i attains the maximum in (5.9) corresponding to x_i and $0 \leq a_i \leq x_i$. The continuity of the functions U and f_{n-1} ensures that these maxima are attained. Now let $a = p_1 a_1 + p_2 a_2$ and note that $0 \leq a \leq x$. This particular choice of a need not be optimal for x in (5.9), so we have

$$f_n(x) \geq U(a) + f_{n-1}(\lambda(x - a)).$$

The functions U and f_{n-1} are concave, which means that $U(a) \geq p_1 U(a_1) + p_2 U(a_2)$ and

$$f_{n-1}(\lambda(x - a)) \geq p_1 f_{n-1}(\lambda(x_1 - a_1)) + p_2 f_{n-1}(\lambda(x_2 - a_2)),$$

since $\lambda(x - a) = p_1\lambda(x_1 - a_1) + p_2\lambda(x_2 - a_2)$. By combining the last three inequalities, we obtain the required result (5.10). $\qquad\square$

The next two examples are concerned with changes in the model we have been using for maximizing total utilities. The first modification is to introduce a discount factor. The idea is that future rewards should be reduced in value because they are not available now. The discount factor θ is fixed in advance, $0 < \theta < 1$, and utilities u_k realized after waiting for k units of time are assigned the present values $\theta^k u_k$, $k = 1, 2, \ldots$. The effect on our previous model is to replace the optimality equation (5.9) by

$$f_n(x) = \max_{0 \leq a \leq x} \{U(a) + \theta f_{n-1}(\lambda(x - a))\}. \tag{5.11}$$

Thus, an immediate utility $U(a)$ is generated by spending an amount a and then state $y = \lambda(x - a)$ is reached after a unit delay. The principle of optimality shows that the maximum future utility at that time is $f_{n-1}(y)$, but this must be discounted because of the delay, so it is equivalent to a present value $\theta f_{n-1}(y)$.

The effect of discounting can be illustrated by considering a modified version of Example 2.2.

Example 5.5 Find the maximum total discounted utility $f_n(x)$ over n periods when the initial capital is x, given that $U(a) = a^{\frac{1}{2}}$, $a \geq 0$.

In the original form of this problem with $\theta = 1$, we obtained the formula $f_n(x) = (c_n x)^{\frac{1}{2}}$, where c_n is a positive constant. A similar formula applies in the discounted case and we shall verify this by induction. We have $f_1(x) = U(x) = x^{\frac{1}{2}}$, so $c_1 = 1$. Now suppose that, for some $n \geq 2$,

$$f_{n-1}(y) = (c_{n-1}y)^{\frac{1}{2}}, \qquad y \geq 0,$$

where c_{n-1} is a positive constant. According to (5.11),

$$f_n(x) = \max_{0 \leq a \leq x} \left\{a^{\frac{1}{2}} + \theta f_{n-1}(y)\right\}$$

with $y = \lambda (x - a)$, so we obtain

$$f_n (x) = \max_{0 \le a \le x} \left\{ a^{\frac{1}{2}} + \left(\theta^2 \lambda c_{n-1}\right)^{\frac{1}{2}} (x - a)^{\frac{1}{2}} \right\}.$$

The maximum occurs when the partial derivative with respect to a vanishes, showing that

$$a = x/ \left(1 + \theta^2 \lambda c_{n-1}\right), \quad f_n (x) = \left(1 + \theta^2 \lambda c_{n-1}\right)^{\frac{1}{2}} x^{\frac{1}{2}}.$$

We may conclude that

$$f_n (x) = (c_n x)^{\frac{1}{2}}$$

always holds, provided that c_n is defined by the relation $c_n = 1 + \theta^2 \lambda c_{n-1}$. Using the fact that $c_1 = 1$, this relation shows that $c_2 = 1 + \theta^2 \lambda$, and so on. The general formula is

$$c_n = 1 + \theta^2 \lambda + \left(\theta^2 \lambda\right)^2 + \cdots + \left(\theta^2 \lambda\right)^{n-1}.$$

Note that the optimal choice of a in the above calculation of $f_n(x)$ is $a = x/c_n$, which defines the decision function $d_n(x) = x/c_n$. We observe that c_n has a finite limit as $n \to \infty$, if $\theta^2 \lambda < 1$. In this case $c_n \to 1/ \left(1 - \theta^2 \lambda\right)$, and

$$f_n (x) \to \left(1 - \theta^2 \lambda\right)^{-\frac{1}{2}} x^{\frac{1}{2}}, \qquad d_n (x) \to \left(1 - \theta^2 \lambda\right) x.$$

The optimal policy in the long run is to spend a proportion $\left(1 - \theta^2 \lambda\right)$ of the available capital at every stage.

Finally, let us consider introducing a regular salary s, instead of the ability to increase capital by earning interest on savings. In this variant of the model, we set $\lambda = 1$ and $\theta = 1$, so there is no discounting of future utilities.

Example 5.6 Suppose that, in each period, the choice of a is restricted by the initial capital x to $0 \le a \le x$ and there is a fixed income s, so that the level of capital is $x - a + s$ at the start of the next period. The utility of spending in any period is measured by a continuous function U with $U(0) = 0$, $U'(a) > 0$ and $U''(a) < 0$ for all $a > 0$.

The conditions guarantee that $U(a)$ is strictly increasing in a and strictly concave. As we shall see, the optimal policy has a simple form and the maximum total utility $f_n(x)$ over n periods can be evaluated explicitly. It will be proved that

$$f_n (x) = U(x) + (n - 1) U(s) \qquad (0 \le x \le s), \qquad (5.12)$$

$$f_n (x) = nU \left(\frac{x + (n - 1) s}{n}\right) \qquad (x \ge s).$$

Proof When $n = 1$, both formulae reduce to $f_1(x) = U(x)$, which is true. Now assume that

$$f_{n-1}(y) = (n-1)U\left(\frac{y + (n-2)s}{n-1}\right),$$

for some $n \geq 2$ and all $y \geq s$. Consider $f_n(x)$ and note that, for any initial state $x \geq 0$ and any choice of a with $0 \leq a \leq x$, the next state $y = x - a + s$ always satisfies the condition $y \geq s$. The optimality equation gives

$$f_n(x) = \max_{0 \leq a \leq x} \{U(a) + f_{n-1}(y)\}$$

and we can use the expression assumed for $f_{n-1}(y)$ with $y = x - a + s$, to obtain

$$f_n(x) = \max_{0 \leq a \leq x} \left\{U(a) + (n-1)U\left(\frac{x - a + (n-1)s}{n-1}\right)\right\}.$$

We have a strictly concave function of a on the right and its derivative with respect to a is

$$U'(a) - U'\left(\frac{x - a + (n-1)s}{n-1}\right),$$

which is strictly decreasing in a. The required maximum can be located by applying Proposition 5.4, parts (ii) and (iii). The derivative has a unique zero when

$$a = \frac{x - a + (n-1)s}{(n-1)}, \qquad a = \frac{x + (n-1)s}{n}. \tag{5.13}$$

The condition that $0 \leq a \leq x$ is satisfied by this choice of a, provided that $x \geq s$. It follows that this is the optimal choice of a if $x \geq s$ and, by using the two equivalent forms of (5.13),

$$f_n(x) = nU(a) = nU\left(\frac{x + (n-1)s}{n}\right).$$

It is clear from this formula that the optimal policy is to spend equal amounts in each of the n periods. We have established that (5.12) is valid whenever $x \geq s$. When $0 \leq x < s$, (5.13) yields a value $a > x$ and this is inadmissible. In this case, we can use the fact that we are maximizing a strictly concave function whose derivative at $a = x$ is $U'(x) - U'(s)$, which is positive if $x < s$. Hence, the maximum must occur at $a = x$, and we obtain

$$f_n(x) = U(x) + (n-1)U(s),$$

as required. Note that this formula corresponds to a spending policy over n periods in which

$$a_1 = x, \qquad a_2 = a_3 = \cdots = a_n = s.$$

The amounts are equal after the first period. $\square$

EXERCISES

5.1 Let $h_j(x), j = 1, 2, \ldots, n$, be convex functions of x. Show that $\sum h_j(x)$ and $\max\{h_1(x), h_2(x), \ldots, h_n(x)\}$ are also convex functions.

5.2 Let h be a convex function defined on the closed interval $[a, b]$ and suppose that $a < x_1 < x_2 < b$. Use the method of Proposition 5.3 to prove that

$$h'(x_1+) \le h'(x_2-).$$

5.3 Suppose that the function h has a second derivative $h''(x)$ at every point. Prove that h is convex if $h''(x) \ge 0$ always and strictly convex if $h''(x) > 0$ for all x.

5.4 Consider the allocation problem of Example 5.3 and let $p_j = 2j$, $c_j = j^2$, for $j = 1, 2, \ldots, N$. Show that the total return is maximized by allocating the whole quantity A to customer k, where k is chosen as the integer nearest to A in the set $\{1, 2, \ldots, N\}$.

5.5 Let $g(a)$ be a differentiable concave function and let $c_1, c_2, \ldots$ be any positive constants. Use a dynamic programming argument to maximize

$$\sum_{j=1}^{n} c_j g(a_j), \qquad \text{subject to} \sum_{j=1}^{n} c_j a_j = x.$$

Deduce Jensen's inequality:

$$\frac{\sum c_j g(a_j)}{\sum c_j} \le g\left(\frac{\sum c_j a_j}{\sum c_j}\right)$$

for any $n \ge 1$ and any $a_1, a_2, \ldots, a_n$.

5.6 A merchant has a quantity x of material to distribute between n customers. His reward for allocating an amount a to customer k is

$$k\left\{\frac{a}{k+1} - c\right\} \quad \text{if } a > 0$$

and zero if $a = 0$, where c is a positive constant. Show that there is an allocation rule for which the total reward is positive if $x > 2c$ and determine an optimal allocation rule when this condition is satisfied.

5.7 A woman has capital $x_0 > 0$ which she wishes to spend and invest in such a way as to maximize her total discounted utility over a specified number of years. At the beginning of each year, she must decide how much of her current capital x to spend. The utility for spending an amount a, $0 < a \leq x$, is given by the natural logarithm, $\log a$. The value of her capital at the beginning of the following year, taking interest into account is $\lambda (x - a)$ where $\lambda > 1$. Future utilities are discounted at the rate θ per year, $0 < \theta < 1$. Let $f_n (x)$ be the maximum total discounted utility over n years, starting with capital x, so $f_1 (x) = \log x$ for $x > 0$.

(i) Obtain the optimality equation for $f_n (x)$.

(ii) Show that $f_n (x) = b_n \log x + c_n$, where b_n and c_n are constants, $n \geq 1$.

(iii) Evaluate b_n for each n, but not c_n, and deduce that the optimal policy is to spend a proportion $(1 - \theta) / (1 - \theta^n)$ of the remaining capital when there are n years left.

Part II

Stochastic Models

6

Markov Systems

6.1 INTRODUCTION

So far, we have been concerned with deterministic models in which the state x changes in a predictable way when we choose a particular action a. Chance plays no part in such models and, once we have decided on a sequence of actions for a given initial state, the state transitions and the corresponding sequence of costs or rewards are determined exactly. The principles of dynamic programming can be extended to stochastic models by allowing for state transitions which involve random variables. This complicates matters because there are many different ways of describing random transitions, but the only fundamental change is that, from now on, we shall be minimizing expected costs or maximizing expected rewards. At each stage, we try to find actions which improve our expectations as far as possible, in the face of an uncertain future.

The state of a dynamic system is basic because it allows us to describe the motion in a convenient way. At time t, the state x_t is a known vector or scalar which provides sufficient information for decisions affecting the future motion of the system.

If we are concerned with future behaviour, a knowledge of the present state allows us to forget the past. The idea of separating past and future is the fundamental characteristic of Markov systems. We shall be concerned with sequential decisions for different types of random processes. They all have the property that, given the present state x_t and action a_t, the probability law $\mathcal{L}$ which determines the probability distribution of the next state x_{t+1}, depends only on x_t, a_t and the time t.

Section 6.2 describes an extension of the general model of Chapter 2 to Markov systems with random jumps from one state to the next. The aim is to minimize the total expected cost over a period and the method of backwards induction can still be applied by considering the conditional expectation of future costs, given the information which becomes available at each decision time.

The next section gives examples to illustrate the new stochastic model and the first is a modified version of Example 2.3 which leads to a straightforward extension of the previous results. The other examples describe simple gambling problems. In Chapters 7 and 8, we shall investigate some optimal stopping problems for Markov processes. The aim is to observe the random transitions of the underlying process and reach a decision to stop it when this is most advantageous. There are many interesting problems of this type.

6.2   STOCHASTIC DYNAMIC PROGRAMMING

Suppose we have a dynamic system with states $x_0, x_1, x_2, \ldots$ at times $t = 0, 1, 2, \ldots$. The motion is controlled by a sequence of actions $a_0, a_1, a_2, \ldots$, in the sense that, given the state x_t at time t, the choice of a_t determines the probability distribution of the next state x_{t+1}. The information available when a_t is chosen includes the previous states and actions : $x_0, a_0, x_1, a_1, \ldots x_{t-1}, a_{t-1}$, but because we are dealing with a Markov system, only the present state x_t is important. The conditional distribution of the next state, given x_t and a_t at time t, is denoted by $\mathcal{L}\left(x_{t+1} \,|\, x_t, a_t, t\right)$. It is helpful to think of the next state as a random variable Y with the probability law $\mathcal{L}\left(y \,|\, x_t, a_t, t\right)$. In practice, this is specified by discrete probabilities or by a probability density function over the range of possible values y. At time $t+1$, the actual value is observed: the random variable Y is replaced by a new state $y = x_{t+1}$, and we can then reconsider the future behaviour of the dynamic system, conditional on x_{t+1} and the choice of a_{t+1}.

The state and action variables may be discrete or continuous and the range of choice of any action may depend on the current state: for simplicity, this is not shown in the notation. The essential characteristics of the model are that, given x_t at time t, the choice of a determines both the probability distribution of the next state x_{t+1} and the expected cost of the transition $x_t \to x_{t+1}$.

The cost c_t of the transition may depend on x_t, a_t and x_{t+1}, but it is convenient to replace the actual cost by its conditional expectation, given the information available at time t. Suppose that the actual cost is $Q\left(x_t, x_{t+1}, a_t, t\right)$, where Q is a prescribed function. This is replaced by its expectation with respect to x_{t+1}. Let us define

$$K\left(x_t, a_t, t\right) = E\left\{Q\left(x_t, Y, a_t, t\right) \,|\, x_t, a_t\right\},   \tag{6.1}$$

where Y has the probability distribution represented by $\mathcal{L}\left(y \,|\, x_t, a_t, t\right)$. Thus, transition costs are specified by a known function K:

$$c_t = K\left(x_t, a_t, t\right),   \tag{6.2}$$

as in the deterministic model of Chapter 2. The difference here is that we need to rely on the fundamental properties of conditional expectations in order to formulate the problem of minimizing the total expected cost over a period.

The objective is to minimize the expectation of the total cost

$$c_0 + c_1 + \cdots + c_{T-1}$$

for a given initial state x_0, by choosing a sequence of actions $a_0, a_1, \ldots, a_{T-1}$. The initial action a_0 will depend on x_0 but, for $t \geq 1$, we must wait until the random process reaches a particular state x_t before we can fix on a definite action $a_t = a_t(x_t, t)$. We cannot know the optimal actions in advance, but we can anticipate by expressing each one in terms of the relevant state. The expected cost is evaluated conditionally, in the following way:

$$E\{c_0 + c_1 + c_2 + \cdots | x_0, a_0\}$$
$$= E\{c_0 + E(c_1 | x_1, a_1) + E(c_2 | x_2, a_2) + \cdots | x_0, a_0\}.$$

Consider a policy π which specifies that, $a_t = \pi_t(x_t)$, for $t = 0, 1, \ldots, T-1$. We have

$$E(c_t | x_t, a_t) = K(x_t, a_t, t). \tag{6.3}$$

Assuming that all the expected costs are finite, the properties of conditional expectations allow us to express the total expectation of the policy π in the form

$$E\left\{\sum_{t=0}^{T-1} c_t \,\middle|\, x_0, a_0\right\} = \sum_{t=0}^{T-1} E\{K(x_t, a_t, t) | x_0, a_0\}. \tag{6.4}$$

The notation here is cumbersome, but we can simplify it by using the time variable

$$n = T - t$$

and the method of backwards induction. The key idea is that, at time t, we are concerned with minimizing the expectation of future costs

$$E\{c_t + c_{t+1} + \cdots c_{T-1} | x_t, a_t\}$$

with respect to the n actions $a_t, a_{t+1}, \ldots, a_{T-1}$. If we can attain these minima for any possible states x_t at times $t \geq 1$, then the total expected cost (6.4) can be minimized. The method is based on the sequence of minimum future expected cost functions.

Definition 6.1 *For any state x and positive integer $n = T - t$, let*

$$f_n(x) = \inf_{a_t, a_{t+1}, \ldots, a_{T-1}} E\{c_t + c_{t+1} + \cdots + c_{T-1} | x_t = x\}.$$

This expresses the minimum future expectation in terms of the state $x_t = x$ at time $t = T - n$. The notation reflects the fact that x and n are the relevant variables.

Consider the future prospects when the present state is x and n is the time to go. If $n = 1$, there is only one transition and we have

$$f_1(x) = \inf_a K(x, a, T - 1).$$

For $n \geq 2$, we need to apply the principle of optimality. The first transition $x \to y$ depends on the choice of an action a but, once the new state y is observed, the effective cost associated with the remaining $n - 1$ transitions is given by $f_{n-1}(y)$. The optimality equation allows for a random jump $x \to Y$, where the distribution of Y is according to $\mathcal{L}(y|x, a, T - n)$. Thus,

$$f_n(x) = \inf_a \left[K(x, a, T - n) + E\{f_{n-1}(Y)|x, a\} \right]. \qquad (6.5)$$

An expectation with respect to Y is also implicit in the first term on the right, but this is included in equation (6.1) which defines the function K. An optimal choice of a in (6.5) yields the corresponding decision function: $a = d_n(x)$. If we can determine the functions $f_1, f_2, \ldots, f_T$ and also a sequence of decision functions $d_1, d_2, \ldots, d_T$, then it should not be difficult to construct an optimal policy which attains the required minimum represented by $f_T(x_0)$.

6.3 APPLICATIONS

We start with a control problem which is a generalization of Example 2.3. Here, the state transitions are affected by random noise but, since the control is linear and costs are quadratic, this produces only minor changes in the results.

Example 6.1 The aim is to control the sequence $\{x_0, x_1, \ldots x_T\}$ so as to minimize the expected sum of all control costs and a terminal cost representing the penalty for missing a target at the point zero. The cost functions K and f_0 are the same as in the deterministic model of Example 2.3.

$$K(x_t, a_t, t) = ca_t^2, \quad f_0(x) = mx^2, \qquad (6.6)$$

where c and m are positive constants. The law of motion is given by the equation

$$x_{t+1} = x_t + a_t + w_t. \qquad (6.7)$$

Each jump $a_t + w_t$ is the sum of the chosen action a_t and a random deviation w_t. We will assume that the random variables $w_0, w_1, \ldots, w_{T-1}$ are independent of one another and that each has the same distribution with known mean μ and variance σ^2. As we shall see, since all the costs are quadratic, the functions $f_1, f_2, \ldots$ depend only on the mean and variance of this distribution. Thus, the probability law $\mathcal{L}$ for a typical transition is defined by

$$Y = x + a + W \qquad (6.8)$$

and the new state Y has

$$E(Y) = x + a + \mu, \qquad \text{var}(Y) = \sigma^2.$$

The calculations begin with

$$f_1(x) = \min_a \left[K(x, a, T - 1) + E\{f_0(Y) | x, a\} \right].$$

Then, by using (6.6) and (6.8), we obtain

$$f_1(x) = \min_a \left[ca^2 + m\left\{ (x + a + \mu)^2 + \sigma^2 \right\} \right].$$

The minimum is found by differentiation, which shows that

$$a = -m(c + m)^{-1}(x + \mu),$$

and hence

$$f_1(x) = cm(c + m)^{-1}(x + \mu)^2 + m\sigma^2.$$

We shall prove by induction that, in general,

$$f_n(x) = m_n(x + n\mu)^2 + (m_0 + m_1 + \cdots + m_{n-1})\sigma^2, \qquad (6.9)$$

where $m_n = cm(c + nm)^{-1}$.

Proof The formulae for f_n and m_n are valid when $n = 1$, taking $m_0 = m$. Now suppose that $f_{n-1}(Y)$ is given by the formula corresponding to (6.9) for some $n \geq 2$. Then we can apply the dynamic programming equation (6.5):

$$f_n(x) = \min_a \left[ca^2 + E\{f_{n-1}(Y) | x, a\} \right].$$

By using the formula for $f_{n-1}(Y)$, we obtain

$$\begin{aligned} E\{f_{n-1}(Y) | x, a\} &= E\left\{ m_{n-1}(Y + (n-1)\mu)^2 \right\} + (m_0 + \cdots + m_{n-2})\sigma^2 \\ &= m_{n-1}(x + a + n\mu)^2 + (m_0 + m_1 + \cdots + m_{n-1})\sigma^2. \end{aligned}$$

The appropriate choice of a is found by minimizing $ca^2 + m_{n-1}(x + a + n\mu)^2$, which leads to

$$a = -m_{n-1}(c + m_{n-1})^{-1}(x + n\mu).$$

It is easy to check that $m_{n-1} (c + m_{n-1})^{-1} = m (c + nm)^{-1}$, so the optimal control $a = d_n (x)$ is given by

$$d_n (x) = -m (c + nm)^{-1} (x + n\mu) . \qquad (6.10)$$

Finally, (6.9) can be verified by collecting together the terms in $f_n (x)$ with this choice of a. □

The decision function d_n in (6.10) does not depend on the variance parameter σ^2. This is an example of *certainty equivalence*: the same optimal control policy can be obtained by assuming a deterministic law of motion in which each random deviation in (6.7) is replaced by its mean: $w_t = \mu$. Note that $d_n (x) \to -\mu$ as $n \to \infty$. The minimum expected cost function given by (6.9) includes a term involving σ^2. When n is large, $f_n(x)$ is dominated by the first term, unless $\mu = 0$, and this is of order $cn\mu^2$. The second term in (6.9) grows like the natural logarithm of n. We have

$$m_0 + m_1 + \cdots + m_{n-1} = m \left\{ 1 + \frac{c}{c + m} + \cdots + \frac{c}{c + (n - 1) m} \right\} ,$$

and this is of order $c \log n$.

There are many applications of stochastic dynamic programming based on gambling games, where the control variable is the size of the stake. Gambling problems are often easy to state, but much more difficult to solve. The next example is rather artificial, but it turns out to have a simple solution.

Example 6.2 In each play of a game, a gambler can bet any non-negative amount up to his current fortune and he will either win or lose that amount with probabilities p and $q = 1-p$, respectively. He is allowed to make n bets in succession, and his objective is to maximize the expectation of the logarithm of this final fortune.

Here, we are maximizing expectations and the terminal reward function is

$$f_0 (x) = \log x,$$

defined for $x > 0$. For $n \geq 1$, since there is no transition cost, relation (6.5) reduces to

$$f_n (x) = \sup_a E \{ f_{n-1} (Y) \,|\, x, a \} .$$

Given the current state $x > 0$ and any choice of a such that $0 \leq a < x$, the new state Y is either $x + a$ with probability p, or $x - a$ with probability q. Hence, the dynamic programming equation is

$$f_n (x) = \sup_{0 \leq a < x} \{ p f_{n-1} (x + a) + q f_{n-1} (x - a) \} . \qquad (6.11)$$

Suppose first that $0 < p \leq \frac{1}{2}$ and $q = 1 - p \geq \frac{1}{2}$, so that the game is unfavourable to the gambler. In this case, it is easy to see that the optimal strategy is always to bet zero. We have

$$f_1(x) = \sup_{0 \leq a < x} \{p \log(x + a) + q \log(x - a)\}. \tag{6.12}$$

The partial derivative $\frac{\partial}{\partial a} \{\cdot\}$ of the expression on the right is

$$\frac{\partial}{\partial a} \{\cdot\} = \frac{p}{x + a} - \frac{q}{x - a}, \tag{6.13}$$

and this is never positive since $p \leq q$ and $p(x + a)^{-1} \leq q(x - a)^{-1}$ when $a \geq 0$. It follows that the maximum in (6.12) is attained by choosing $a = 0$ and $f_1(x) = p \log x + q \log x = \log x$. Thus, $f_1(x) = f_0(x)$ for all $x > 0$. It follows immediately from (6.11) that $f_2(x) = f_1(x)$ and so on. Hence,

$$f_n(x) = \log x$$

for $n \geq 0$, $x > 0$ and the optimal strategy always corresponds to $a = 0$: never gamble.

Now consider the case, $\frac{1}{2} < p < 1$, $p > q$, when the game is favourable. Equations (6.12) and (6.13) still apply and an examination of the second derivative with respect to a shows that the expression on the right of (6.12) is strictly concave in a. The required maximum occurs when

$$\frac{p}{x + a} = \frac{q}{x - a},$$

and this determines

$$a = (p - q)x. \tag{6.14}$$

When this optimal stake is substituted in (6.12), it leads to

$$f_1(x) = \log x + p \log p + q \log q + \log 2.$$

Thus, $f_1(x)$ differs from $f_0(x)$ by a constant. Then it is a straightforward matter to prove that

$$f_n(x) = \log x + ng, \tag{6.15}$$

where $g = p \log p + q \log q + \log 2$.

In this example, the optimal choice of action does not depend on n and (6.14) holds for all $n \geq 1$ and $x > 0$. provided that $p > q$.

We now turn to a gambling problem where the aim is to reach a specified fortune s before ruin occurs.

Example 6.3 Imagine that you have a small amount of money and that you need to acquire a total s, before losing all of it, by gambling on the toss of a coin. If you have an amount j at any stage, you may bet $a = 1, 2, \ldots, j$ on the next toss so that j is replaced by $j+a$ or $j-a$ with probabilities p and $q = 1-p$, respectively. The aim is to maximize the probability of reaching the terminal state s before the random process is absorbed at state zero, representing ruin. Let us confine our attention to strategies which avoid overshooting the target by choosing a so that $j + a \leq s$. Thus, given $j = 1, 2, \ldots, s - 1$, the amount a gambled on the next toss of the coin is restricted to

$$a = 1, 2, \ldots, \min(j, s - j). \tag{6.16}$$

The optimal strategy depends critically on whether the coin-tossing game is favourable or not. The simplest case is when the coin is fair: $p = q = \frac{1}{2}$. Then any strategy satisfying condition (6.16) is optimal.

If $p > \frac{1}{2}$ and $p > q$, the optimal procedure is to choose $a = 1$ always, as we shall see. The unfavourable case, $0 < p < \frac{1}{2}$ is more difficult to analyse, except when the target s is small. Examples suggest that when the underlying game is unfavourable, it is best to stake as much as possible, choosing $a = \min(j, s-j)$ when the amount available is j.

Let f_j be the maximum probability of success, starting at j. In other words, f_j is the probability of reaching state s before state zero, under an optimal strategy. Clearly $f_0 = 0$ and $f_s = 1$. For $j = 1, 2, \ldots, s - 1$, the principle of optimality shows that

$$f_j = \max_a \{pf_{j+a} + qf_{j-a}\}, \tag{6.17}$$

where the amount a bet on the first toss is subject to condition (6.16).

Suppose that the coin is fair. Then it is easy to verify that the solution is defined by

$$f_j = \frac{j}{s}, \tag{6.18}$$

$j = 0, 1, \ldots, s$. Notice that this is linear in j and

$$\tfrac{1}{2}f_{j+a} + \tfrac{1}{2}f_{j-a} = f_j$$

holds for any admissible choice of a. It follows that

$$f_j = \max_a \left\{ \tfrac{1}{2}f_{j+a} + \tfrac{1}{2}f_{j-a} \right\}.$$

It can be shown that any solution of the optimality equation (6.17) is unique: see Exercise 6.3.

Now suppose that $\frac{1}{2} < p < 1$ and consider the policy of always choosing $a = 1$. If this is optimal we must have

$$f_j = pf_{j+1} + qf_{j-1}$$

when $0 < j < s$ and $f_0 = 0$, $f_s = 1$. Since $p + q = 1$, an equivalent relation is

$$p\{f_{j+1} - f_j\} = q\{f_j - f_{j-1}\}, \qquad (6.19)$$

and it is a straightforward exercise to apply the boundary conditions at $j = 0$ and $j = s$ and establish that

$$f_j = \frac{p^s - p^{s-j}q^j}{p^s - q^s}. \qquad (6.20)$$

We shall prove that this formula also gives the solution of the optimality equation (6.17), showing that the policy of always choosing $a = 1$ is optimal.

Proof This will be done by demonstrating that when f_j is given by (6.20), the expression on the right of (6.17) is decreasing in a. Note that (6.19) holds for $j = 1, 2, \ldots, s - 1$, and by applying this repeatedly we can deduce that

$$p\{f_{j+a} - f_{j+a-1}\} = (q/p)^{2(a-1)} q\{f_{j-a+1} - f_{j-a}\}.$$

This is valid for $a = 1, 2, \ldots, \min(j, s - j)$ and the factor $(q/p)^{2(a-1)} < 1$ for $a \geq 2$, so we have

$$p\{f_{j+a} - f_{j+a-1}\} < q\{f_{j-a+1} - f_{j-a}\}.$$

Finally, this can be rearranged and used to show that

$$pf_{j+a} + qf_{j-a} < pf_{j+a-1} + qf_{j-a+1} < \cdots < pf_{j+1} + qf_{j-1}.$$

Since, $pf_{j+1} + qf_{j-1} = f_j$, it follows that (6.17) holds, with the maximum achieved by $a = 1$. $\qquad \square$

As we noted earlier, the unfavourable case $0 < p < \frac{1}{2}$ is more difficult. The simplest example is when $s = 4$ and this will be used to illustrate the problem. Consider the strategy of choosing $a = \min(j, 4 - j)$, which means that $a = 1$ when $j = 1$ or 3 and $a = 2$ when $j = 2$. If this is the optimal strategy, then equation (6.17) and the boundary conditions show that $f_0 = 0$, $f_4 = 1$ and

$$f_1 = pf_2, \quad f_2 = p, \quad f_3 = p + qf_2. \qquad (6.21)$$

There is also an optimality condition which corresponds to choosing $a = 2$, rather than $a = 1$, when $j = 2$. This requires that

$$f_2 = pf_4 + qf_0 \geq pf_3 + qf_1. \qquad (6.22)$$

The solution of equations (6.21) is given by

$$f_1 = p^2, \quad f_2 = p, \quad f_3 = p(1 + q).$$

Hence, condition (6.22) is equivalent to

$$p \geq p^2 (1 + q) + p^2 q, \quad \text{or} \quad 1 \geq p + 2pq$$

and this is satisfied when $p < \frac{1}{2}$.

The strategy of choosing $a = \min(j, s - j)$ is optimal when $p < \frac{1}{2}$, for any s. A general proof, using a dynamic programming argument, is given in Chapter IV of Ross [Ros83]. The above method which relies on finding an explicit solution of the optimality equation is useful only when s is small: see Exercises 6.5 and 6.6.

EXERCISES

6.1 Consider the control problem of Example 6.1 and the limiting policy of choosing $a = -\mu$ at every stage, suggested by letting $n \to \infty$ in (6.10). Show that this policy leads to an expected cost over n stages given by

$$g_n(x) = nc\mu^2 + m\left(x^2 + n\sigma^2\right)$$

for any initial state x and $n \geq 0$. Compare the long-term average costs obtained as the limits of $g_n(x)/n$ and the minimal ratio $f_n(x)/n$ when $n \to \infty$.

6.2 Prove that the maximum expectation in Example 6.2 is given by equation (6.15), provided that $p > \frac{1}{2}$. Show that the constant

$$g = p \log p + (1 - p) \log (1 - p) + \log 2$$

is strictly positive for $\frac{1}{2} < p < 1$.

6.3 Prove that the optimality equation (6.17) in example 6.3 has at most one solution by considering the difference of two possible solutions.

6.4 Show that the solution of equation (6.19) for $j = 1, 2, \ldots, s - 1$, subject to the boundary conditions $f_0 = 0$ and $f_s = 1$, is given by (6.20) except when $p = q = \frac{1}{2}$.

6.5 Consider the case $0 < p < \frac{1}{2}$ in Example 6.3 with the bold strategy defined by $a = \min(j, s - j)$. Show that this strategy is optimal when $s = 5$ by solving the equations for the probabilities of success $f_j, j = 1, 2, 3, 4$, and verifying the optimality equation (6.17). This requires that the following conditions must be satisfied:

$$f_2 \geq p f_3 + q f_1 \quad \text{and} \quad f_3 \geq p f_4 + q f_2 \quad .$$

6.6 Show that the bold strategy is optimal, as in the previous exercise, when $0 < p < \frac{1}{2}$ and $s = 6$. In this case, there are four optimality conditions to be satisfied. Note that if $s = 8$, the number of optimality conditions rises to nine.

7

Optimal Stopping

7.1 INTRODUCTION

The simplest kind of optimization problem for a random process involves just two possible actions: stop or continue. At each state which occurs, a decision must be made whether to stop the process or to wait for the result of the next transition and then reconsider the position. There are many different problems of this type in which the underlying random process is Markovian. This means that the future behaviour of the process depends only on its present state and not on its previous history. Hence, a decision whether to terminate the process or not can be based on comparing the reward for stopping at the present state with the best expectation that can be achieved after allowing another transition.

This chapter is concerned with optimal stopping for Markov chains and we shall begin with an introduction to such processes, in which both state and time variables are discrete. Optimal stopping theory is developed in the next section, showing how stopping times can be determined by dividing the state space into continuation and stopping sets. Section 7.3 gives several applications. There are well known problems of optimal stopping which do not fit conveniently into the pattern and notation of Markov chains, although the essential ideas are the same. Some of these special problems will be described in Chapter 8.

Consider a Markov chain with possible states, $0, 1, 2, \ldots$ whose motion is governed by a probability law specified by the transition matrix P. Thus, $P = (p_{ij})$ is an infinite matrix with elements $p_{ij} \geq 0$, such that $\sum_j p_{ij} = 1$. For any given state $i \geq 0$, the probabilities p_{ij}, $j = 0, 1, 2, \ldots$, describe the distribution of the next state. We have a Markov system $x_0, x_1, x_2, \ldots$ in which every x_t belongs to the state space $\{0, 1, 2, \ldots\}$ and, given that $x_t = i$ at any time t, the next state is a random variable with the conditional probabilities

$$P\left(x_{t+1} = j \mid x_t = i\right) = p_{ij}. \tag{7.1}$$

The process is homogeneous in time since the transition probabilities do not depend on t and the fact that x_{t+1} must also belong to the state space leads to the condition that $\sum_j p_{ij} = 1$. Matrix notation is useful here because, as we shall see, the transition matrix for n transitions is simply the nth power of P. The distribution of the state x_t when t is large is related to the asymptotic behaviour of P^n as $n \to \infty$. This is described in many books on probability and stochastic processes: see, for example, the classic by Feller [Fel68] or the more recent text by Grimmett and Stirzaker [GS82]. However, we shall not be concerned with the asymptotic theory of Markov chains. Optimal stopping can be investigated by focussing on a typical state and the possibilities arising from a single transition.

The n-step transition matrix $P^{(n)} = \left(p_{ij}^{(n)} \right)$ is made up of the transition probabilities over n units of time:

$$p_{ij}^{(n)} = P\left(x_{t+n} = j \,|\, x_t = i\right)$$

for each $n \geq 1$ and, of course, $P^{(1)} = P$. These transition probabilities are related by the Chapman–Kolmogorov equations:

$$p_{ik}^{(m+n)} = \sum_j p_{ij}^{(m)} p_{jk}^{(n)}. \tag{7.2}$$

We can establish this by considering any transition $i \to k$ over $m + n$ steps and conditioning on the state j reached after m steps.

$$
\begin{aligned}
p_{ik}^{(m+n)} &= P\left(x_{m+n} = k \,|\, x_0 = i\right) \\
&= \sum_j P\left(x_{m+n} = k, x_m = j \,|\, x_0 = i\right) \\
&= \sum_j P\left(x_{m+n} = k \,|\, x_m = j, x_0 = i\right) P\left(x_m = j \,|\, x_0 = i\right).
\end{aligned}
$$

The Markov property shows that the first conditional probabilities on the right do not depend on the initial state i, so they reduce to $p_{jk}^{(n)}$. Hence, the equation is equivalent to (7.2).

There is a natural extension of matrix multiplication to infinite matrices and, by using this, we can replace the set of Chapman–Kolmogorov equations for all $i, k \geq 0$ by the matrix equation

$$P^{(m+n)} = P^{(m)} P^{(n)}$$

This holds for any integers $m, n \geq 1$ and it follows that $P^{(n)} = P^n$, the nth power of P.

We now introduce a structure of rewards and costs in order to specify an optimal stopping problem. The underlying random process is the Markov chain with states $0, 1, 2, \ldots$, and one-step transition matrix $P = (p_{ij})$, where $p_{ij} \geq 0$ and $\sum_j p_{ij} = 1$ for each $i \geq 0$. Suppose that the reward for stopping at state i is $r_i \geq 0$. This is a terminal reward in the sense that a decision to stop when $x_t = i$ is final and no further costs or rewards can occur. The immediate cost associated with a decision to continue from state i is $c_i \geq 0$, which represents the cost of the next transition only. It can be interpreted as the expected cost of replacing $x_t = i$ by a new state $x_{t+1} = j$ described by the probability distribution p_{ij}, $j = 0, 1, 2, \ldots$.

Let v_i be the maximum expected reward net of costs, given an initial state i. For the moment, let us assume that an optimal stopping time exists for each $i \geq 0$ and that v_i is the corresponding expectation, including all transition costs as well as the terminal reward. Note that $v_i \geq r_i$, since it is permissible to stop the random process at its initial state $x_0 = i$. The alternative is to allow one transition $i \rightarrow j$, at least, and this should produce a net expectation $-c_i + \sum_j p_{ij} v_j$, according to the principle of optimality. The decision whether to stop or continue must depend on whether the reward r_i for stopping exceeds this or not, which leads to the optimality equation

$$v_i = \max\left\{ r_i, -c_i + \sum_j p_{ij} v_j \right\}. \tag{7.3}$$

The function $\{v_i, \ i = 0, 1, 2, \ldots\}$ is called the value function. It will be shown in the next section, under suitable conditions, that it satisfies equation (7.3) and that an optimal stopping time is determined by the rule:

$$\text{stop at } x_t = i \text{ if } v_i = r_i, \text{ continue if } v_i > r_i.$$

The following simple example indicates some of the possibilities.

Example 7.1 Consider a process known as the simple random walk on the integers $0, 1, 2, \ldots$. This is defined by the following transition probabilities. For $i \geq 1$,

$$p_{i,i+1} = p_{i,i-1} = \tfrac{1}{2},$$

and for $i = 0$, $p_{00} = p_{01} = \tfrac{1}{2}$. The transition matrix is

$$P = \begin{pmatrix} \tfrac{1}{2} & \tfrac{1}{2} & \cdot & \cdot & \cdot & \cdot \\ \tfrac{1}{2} & \cdot & \tfrac{1}{2} & \cdot & \cdot & \cdot \\ \cdot & \tfrac{1}{2} & \cdot & \tfrac{1}{2} & \cdot & \cdot \\ \cdot & \cdot & \tfrac{1}{2} & \cdot & \tfrac{1}{2} & \cdot \end{pmatrix}$$

in which all the $\cdot$ entries are zero. Another way of describing the transitions is to write $i \rightarrow i \pm 1$, each with probability $\tfrac{1}{2}$, if $i > 0$. The state $i = 0$ acts as a barrier since $0 \rightarrow 1$ and $0 \rightarrow 0$ are the only possible transitions from it.

Let $c_i = 1$ for all $i \geq 0$ and suppose that the rewards for stopping are given by $r_i = 2$ if i is an even integer, including zero, and $r_i = 0$ if i is odd. In this case, the solution is obvious. The maximum possible expectation is 2, so it is optimal to stop at any even state and $v_i = 2$. If i is odd, a single transition must lead to an even state. The optimal stopping time is $T = 1$, and this produces a net expectation of $v_i = -1 + 2 = 1$. It is easily verified that the value function given by

$$v_i = 2, \quad i = 0, 2, \ldots, \text{ and } v_i = 1, \quad i = 1, 3, \ldots$$

satisfies the optimality equation (7.3).

A different specification of the costs and rewards will illustrate the fact that any subset of the state space can serve as the optimal stopping set. Let S be a non-empty set contained in the state space and let C be its complement. Now suppose that $c_i = 0$ for all $i \geq 0$ and define

$$r_i = 0 \text{ if } i \in C, \qquad r_i = 1 \text{ if } i \in S.$$

Since there is no cost for allowing the random process to continue, it is clearly advantageous to wait until it reaches a state in S where the reward is 1. Note that this is not the only optimal stopping rule. Consider a single stopping point $s \in S$. The random walk has the property that it will eventually reach a particular state s from any initial state i. The policy of waiting until s is reached is also optimal.

7.2 STOPPING TIMES AND STOPPING SETS

Let us return to the general problem of optimal stopping for a Markov chain. Its motion is determined by a transition matrix $P = (p_{ij})$ with entries p_{ij} for $i, j, = 0, 1, 2, \ldots$. This must be a *stochastic* matrix which, by definition, satisfies the conditions

$$p_{ij} \geq 0 \text{ and } \sum_j p_{ij} = 1, \qquad i = 0, 1, 2, \ldots.$$

The reward for stopping at state i is r_i and the continuation cost for the next transition is c_i.

A stopping time T is a non-negative random variable associated with the Markov chain $\{x_0, x_1, x_2, \ldots\}$. It may depend on the sequence of states, but it must never anticipate the future. For example, it is inadmissible to stop when the sequence of rewards $\{r_{x_t}, t = 0, 1, 2, \ldots\}$ reaches its maximum, because this would imply anticipating that no higher reward is achieved at a later time. Roughly speaking, a stopping time T must have the property that, for any $t \geq 0$, the event $\{T = t\}$ is determined by the sequence of states $\{x_0, x_1, \ldots, x_t\}$. In other words, whether the random process stops at time t or not depends only on its history up to the time t. We shall be mainly concerned with stopping times based on stopping sets, in the following way. Let S be a subset of the state space and suppose that the initial state of the

Markov chain is $x_0 = i$. The corresponding stopping time T is defined as the first time that the process reaches a state in S:

$$T = \inf \{t \geq 0 : x_t \in S\}.$$

Thus $T = 0$ if $i \in S$. If $x_t \notin S$ for all $t \geq 0$, we write $T = \infty$.

Consider the expected net reward, starting at state i and using a stopping time T. There is a cost for each transition, so we obtain the conditional expectation

$$E\left\{-c_{x_0} - c_{x_1} - \cdots - c_{x_{T-1}} + r_{x_T} \mid x_0 = i\right\}.$$

The maximum expected net reward v_i is defined by

$$v_i = \sup_{T \geq 0} E\left\{-c_{x_0} - c_{x_1} - \cdots - c_{x_{T-1}} + r_{x_T} \mid x_0 = i\right\}. \tag{7.4}$$

A monotone sequence of approximations
Imagine that we are permitted at most n further steps before stopping. This restricts the choice of stopping times so that $0 \leq T \leq n$, and enables us to use a dynamic programming approach. For each $i \geq 0$ and any non-negative integer n, let $u_i(n)$ be the maximum expected net reward, given the initial state i and allowing at most n transitions before stopping. Clearly,

$$u_i(0) = r_i, \qquad i = 0, 1, 2, \ldots$$

and, for $n \geq 1$, the principle of optimality shows that

$$u_i(n) = \max\left\{r_i, -c_i + \sum_j p_{ij} u_j(n-1)\right\} \tag{7.5}$$

for each $i \geq 0$. The device of introducing an artificial time horizon n leads to a monotone sequence of approximations to the value function $\{v_i, i \geq 0\}$.

It is clear that the maximum expectation $u_i(n)$ is non-decreasing in n, since the class of admissible stopping times becomes larger when n is replaced by $n + 1$. We have

$$u_i(n+1) \geq u_i(n), \qquad i \geq 0, n \geq 0. \tag{7.6}$$

This inequality can be verified easily by using the relation (7.5). Intuitively, it is clear that $u_i(n)$ converges to v_i as $n \to \infty$, if it has a finite limit. We need to make suitable assumptions in order to ensure this. It will be assumed throughout this section that

$$0 \leq r_i \leq b \text{ and } c_i \geq c > 0 \tag{7.7}$$

holds for all $i \geq 0$, where b and c are positive constants. The first assumption ensures that $u_i(n) \leq b$ always and the second excludes infinite stopping times, as we shall see later. These conditions are not strictly necessary for a satisfactory theory, but they will be useful in simplifying the presentation.

Proposition 7.1 *Under the assumption (7.7), for each $i \geq 0$ and all $n \geq 0$,*

$$r_i \leq u_i(n) \leq b \text{ and } u_i(n) \to u_i \leq b$$

as $n \to \infty$. Further, the limit function $\{u_i, i \geq 0\}$ satisfies the optimality equation:

$$u_i = \max\left\{r_i, -c_i + \sum_j p_{ij}u_j\right\}.$$

Proof We have $u_i(0) = r_i \leq b$ and

$$u_i(1) = \max\left\{r_i, -c_i + \sum_j p_{ij}r_j\right\}.$$

Since the costs c_i are positive, the second term on the right here is bounded above by $\sum_j p_{ij}r_j \leq \sum_j p_{ij}b = b$. Hence $r_i \leq u_i(1) \leq b$. Assume, inductively, that $u_j(n-1) \leq b$ for all $j \geq 0$. Then (7.5) shows that $u_i(n) \geq r_i$ and the terms on the right of this equation satisfy $r_i \leq b$ and

$$-c_i + \sum_j p_{ij}u_j(n-1) \leq \sum_j p_{ij}b = b.$$

It follows that $r_i \leq u_i(n) \leq b$ and the induction is complete. Since the sequence $\{u_i(n), n = 0, 1, 2, \ldots\}$ is monotone and bounded, it has a limit $u_i \leq b$.

Finally, we let $n \to \infty$ in relation (7.5). For each $i \geq 0$ and any $k \geq 1$,

$$\sum_{j=0}^{k} p_{ij}u_j(n-1) \to \sum_{j=0}^{k} p_{ij}u_j$$

as $n \to \infty$. The sum on the left is increasing in n and also in k. The sum on the right converges to $\sum_j p_{ij}u_j$ as $k \to \infty$. We may conclude that

$$\sum_{j=0}^{\infty} p_{ij}u_j(n-1) \to \sum_{j=0}^{\infty} p_{ij}u_j$$

as $n \to \infty$. Since relation (7.5) holds for every $n \geq 1$, it ensures that

$$u_i = \max\left\{r_i, -c_i + \sum_j p_{ij}u_j\right\},$$

as required. □

Optimal continuation and stopping sets

The optimality equation (7.3) may have more that one solution. Spurious solutions can arise which do not correspond to proper stopping times. However, the limits u_i, $i \geq 0$, established in Proposition 7.1 will be shown to represent the required maximum expectations. To see this, we shall consider continuation and stopping sets determined by each of the approximations $\{u_i(n), i \geq 0\}$. The limit function $\{u_i, i \geq 0\}$ satisfies the optimality equation. The inequality $u_i \geq r_i$ always holds and, if $u_i > r_i$, there is a definite advantage in allowing the random process to continue further when it reaches state i. This suggests dividing the state space into two decision regions defined by

$$C = \{i \geq 0 : u_i > r_i\}, \qquad S = \{i \geq 0 : u_i = r_i\}.$$

These sets form the optimal continuation and stopping regions for the problem, and it will be shown that the rule:

stop as soon as the process reaches a state in S

determines an optimal stopping time T_i, for any initial state i.

For each $n \geq 1$, let

$$C(n) = \{i \geq 0 : u_i(n) > r_i\}, \qquad S(n) = \{i \geq 0 : u_i(n) = r_i\}.$$

These sets cannot overlap and their union $C(n) \cup S(n)$ covers the state space. The inequality (7.6) shows that we have an increasing sequence of continuation regions: $C(1) \subset C(2) \subset \cdots$. Hence, the stopping regions form a decreasing sequence: $S(1) \supset S(2) \supset \cdots$. Further, since $u_i(n) \leq u_i$ always holds, we have $C(n) \subset C$ and $S(n) \supset S$. Note that a state $i \in C$ if and only if $u_i(n) > r_i$ for sufficiently large values of n, but $i \in S$ if $u_i(n) = r_i$ for all $n \geq 1$.

We recall that $u_i(n)$ is the maximum expected net reward for an initial state i when at most n transitions are allowed. A precise definition is obtained by modifying (7.4) so that the supremum is restricted to stopping times with $0 \leq T \leq n$. However, this gives little indication of how to find a stopping time $T_i(n)$ which attains the maximum expectation $u_i(n)$. A more revealing approach is to investigate relation (7.5). The technique of backwards induction can be used to prove that $u_i(n)$ is attained by the following stopping time:

$$T_i(n) = \min\{t \geq 0 : x_t \in S(n-t) \mid x_0 = i\}. \qquad (7.8)$$

A brief outline of the argument is given below.

The set $S(0)$ is interpreted in (7.8) as the whole state space, so that $x_n \in S(0)$ always holds and, hence $T_i(n) \leq n$. Thus, $T_i(n) = 0$ if $i \in S(n)$, $T_i(n) = 1$ if $i \in C(n)$ is followed by a state $x_1 \in S(n-1)$, and so on. The stopping time $T_i(n)$ is the first time $t \geq 0$ that the Markov chain $\{x_0, x_1, x_2, \ldots\}$ reaches a state $x_t \in S(n-t)$ from the given initial state

$x_0 = i$. The process is stopped when no advantage can be gained by using any of the remaining $n - t$ steps. The stopping rule is complicated by the use of an increasing sequence of stopping sets $S(n) \subset S(n-1) \subset \ldots \subset S(0)$. However, this complication disappears in the limit as $n \to \infty$, because each of these sets is replaced by the optimal stopping set S.

It is easy to see that $u_i(1)$ is achieved by the stopping time $T_i(1)$, for each $i \geq 0$. We have

$$u_i(1) = \max \left\{ r_i, -c_i + \sum_j p_{ij} r_j \right\}$$

and, if $i \in S(1)$ and $T_i(1) = 0$, the corresponding reward is $u_i(1) = r_i$. On the other hand, if $i \in C(1)$ and $T_i(1) = 1$, then $u_i(1) = -c_i + \sum_j p_{ij} r_j$, which is the expected net reward after one transition. Similarly,

$$u_i(2) = \max \left\{ r_i, -c_i + \sum_j p_{ij} u_j(1) \right\},$$

and there are two cases: either $i \in S(2)$ and $T_i(2) = 0$ or $i \in C(2)$ and $T_i(2) \geq 1$. In the first case, $u_i(2) = r_i$, and otherwise $T_i(2)$ takes the values 1 or 2 according to whether the first transition $i \to j$ leads to a state j in $S(1)$ or $C(1)$, respectively. In fact, $T_i(2) = 1 + T_j(1)$ and, since $T_j(1)$ is the stopping time which attains $u_j(1)$, it follows that

$$u_i(2) = -c_i + \sum_j p_{ij} u_j(1)$$

is attained by $T_i(2)$. By repeating the argument for $u_i(3)$ and so on, we can establish the general result that the maximum net expectation $u_i(n)$ is attained by the stopping time $T_i(n)$.

Now consider what happens when $n \to \infty$. For each initial state $i \geq 0$, the sequence of random variables $\{T_i(n), n \geq 1\}$ converges to the random variable T_i. Proposition 7.1 shows that the corresponding sequence of expectation $\{u_i(n), n \geq 1\}$ converges to u_i. To establish that the stopping times T_i are optimal, we need to verify that T_i attains the supremum in (7.4), for any $i \geq 0$. Strictly speaking, a rigorous proof of these claims would involve results in probability theory which are beyond the scope of this book. However, we can use a more elementary approach which indicates the main steps of the proof.

The behaviour of the sequence of random variables $\{T_i(n), n \geq 1\}$ can be demonstrated by considering a fixed realization of the Markov chain. Let $x_0 = i$ and let $\{x_0, x_1, x_2, \ldots\}$ be any possible sequence of states. This determines the corresponding values of $T_i(1), T_i(2)$, etc. according to the definition (7.8). Since $S(n+1-t) \subset S(n-t)$ whenever $0 \leq t \leq n$, it is clear that the values $\{T_i(n), n \geq 1\}$ form a non-decreasing sequence of integers which must converge to a limit and it follows from the definition of the stopping region S that this limit is the value of T_i. As we shall see, infinite values of the limit random variables T_i are effectively excluded because the stopping times have uniformly bounded expectations.

It follows from Proposition 7.1 and assumptions (7.7) that $0 \leq u_i(n) \leq b$ always holds. We also know that

$$u_i(n) = E\left\{-c_{x_0} - c_{x_1} - \cdots - c_{x_{T-1}} + r_{x_T} \mid x_0 = i\right\},$$

where T is the stopping time $T_i(n)$ in (7.8). Since $c_{x_t} \geq c$ for all $t \geq 0$ and $r_{x_t} \leq b$ always holds, we obtain

$$0 \leq u_i(n) \leq b - cE\{T_i(n)\},$$

and it follows that

$$E\{T_i(n)\} \leq \frac{b}{c}. \tag{7.9}$$

This holds for every $i \geq 0$ and $n \geq 1$.

The sequence of random variables $\{T_i(n), n \geq 1\}$ is increasing, in the sense described earlier, and its limit is the stopping time T_i. In particular, for each integer $k \geq 1$, $P(T_i(n) \geq k)$, increases to $P(T_i \geq k)$ as $n \to \infty$. We shall apply two elementary results in probability which hold for any stopping time T. The first is that its expectation can be expressed in the form

$$E\{T\} = \sum_{k=1}^{\infty} P(T \geq k) \tag{7.10}$$

and the second is Markov's inequality:

$$P(T \geq k) \leq \frac{1}{k}E\{T\}.$$

For example, from (7.9) we obtain

$$P(T_i(n) \geq k) \leq \frac{b}{ck}.$$

All these probabilities are increasing as n increases and in the limit as $n \to \infty$, we have

$$P(T_i \geq k) \leq \frac{b}{ck}.$$

This can be used to verify that T_i is finite with probability 1 and we can use the fact that

$$E\{T_i(n)\} = \sum_{k=1}^{\infty} P(T_i(n) \geq k) \leq \frac{b}{c}$$

to obtain a stronger result. Since the terms of this series all increase with n and their sum is uniformly bounded, it follows that

$$E\{T\} \leq \frac{b}{c} \qquad (7.11)$$

for each $i \geq 0$.

The inequality (7.11) is an important step on the way to establishing the optimality of the stopping time T_i. We can rely on the results of Proposition 7.1 and the fact that, for each finite horizon, the maximum expectation $u_i(n)$ is achieved by the corresponding stopping time $T_i(n)$. The techniques already described can be extended to justify the required conclusion, but only after further detailed analysis. The result will be stated without formal proof.

Proposition 7.2 *Suppose that conditions (7.7) hold. For each $i \geq 0$, let $u_i = \lim_{n \to \infty} u_i(n)$. Then $\{u_i, i \geq 0\}$ is the value function defined in (7.4) and each maximum expectation u_i is attained by the stopping time T_i based on the optimal stopping region $S = \{j \geq 0 : u_j = r_j\}$.*

The one-step look-ahead rule

There is a simple policy based on the stopping set $S(1)$. It is known as the One-Step Look-Ahead rule (OSLA). We recall that

$$S(1) = \{i \geq 0 : u_i(1) = r_i\},$$

where $u_i(1)$ is the maximum net expected reward, given an initial state i, when at most one transition is allowed:

$$u_i(1) = \max\left\{ r_i, -c_i + \sum_j p_{ij} r_j \right\}.$$

The one-step look-ahead procedure corresponds to stopping as soon as the underlying Markov chain reaches a state in $S(1)$. In other words, it is defined by the rule

> *OSLA: continue from state i if and only if there is an advantage in taking a single step and then stopping.*

This is equivalent to continuing if and only if

$$r_i < -c_i + \sum_j p_{ij} r_j. \qquad (7.12)$$

The OSLA rule is applied repeatedly at successive states visited by the random process until it reaches $S(1)$. This is not the same as using the stopping times $T_i(1)$ which attain the expectations $u_i(1)$ for each initial state.

In fact, OSLA does better than this. It can be shown that the expectation achieved by the OSLA rule is at least $u_i(1)$, for any $i \geq 0$: see Exercise 7.2. It is a useful sub-optimal procedure which sometimes provides a good policy. Under very special conditions, it determines the optimal policy. As we shall prove, this occurs if the stopping set $S(1)$ is *closed*, in the sense that transitions from $S(1)$ to states outside it are not possible.

Proposition 7.3 *The OSLA rule is optimal for any initial state if the set $S(1)$ is closed: that is, if $p_{ij} = 0$ whenever $i \in S(1)$ and $j \notin S(1)$.*

Proof In general, the stopping regions $S(n), n \geq 1$, form a decreasing sequence and their intersection is the optimal stopping set S. It will be enough to prove that, if $S(1)$ is closed, all these stopping regions coincide and $S(1) = S$. This will be established by showing that, for each $n \geq 1$,

$$u_j(n) = r_j \text{ if } j \in S(1). \qquad (7.13)$$

It is true when $n = 1$, by definition of $S(1)$, so let us assume that (7.13) holds for some integer $n \geq 1$. Then, by relation (7.5),

$$u_i(n+1) = \max\left\{r_i, -c_i + \sum_j p_{ij} u_j(n)\right\}.$$

Now fix $i \in S(1)$ and notice that the sum on the right is equivalent to a sum over $S(1)$:

$$\sum_{j=0}^{\infty} p_{ij} u_j(n) = \sum_{j \in S(1)} p_{ij} u_j(n),$$

since $p_{ij} = 0$ except when $j \in S(1)$. Then our assumption (7.13) means that

$$u_i(n+1) = \max\left\{r_i, -c_i + \sum_j p_{ij} r_j\right\}.$$

Thus, $u_i(n+1) = u_i(1) = r_i$ for each $i \in S(1)$, and this completes the inductive proof of (7.13).

We have shown that, for any state $j \in S(1), u_j(n) = r_j$ holds for $n = 1, 2, \ldots$. Hence,

$$u_j = \lim_{n \to \infty} u_j(n) = r_j,$$

and it follows that $j \in S$. We may conclude that $S(1) \subset S$ and since $S(1) \supset S$ also holds, $S(1) = S$. In other words, $S(1)$ is the optimal stopping set and the OSLA rule determines the optimal stopping times. □

7.3   APPLICATIONS

We shall begin with two examples based on a Markov chain which can only move to the right through the integers. The first shows that, if there is no cost for allowing the process to continue, there may be no optimal policy. The second illustrates the OSLA rule.

Example 7.2 Consider the Markov chain defined by the transition probabilities

$$p_{i,i} = p_{i,i+1} = \tfrac{1}{2} \qquad (i \geq 0). \qquad (7.13)$$

Each transition consists of moving a distance 0 or 1 to the right, with equal probabilities. Suppose that every $c_i = 0$ and that the sequence of rewards is strictly increasing

$$0 \leq r_0 < r_1 < r_2 < \cdots .$$

Condition (7.7) is not satisfied here, since there is no cost for continuation, and it is easy to see that the optimal stopping set is empty, so there is no optimal stopping time.

Notice that the OSLA rule (7.12) is to continue further from a state j if and only if

$$r_j < \tfrac{1}{2}r_j + \tfrac{1}{2}r_{j+1}, \qquad \text{i.e. } r_j < r_{j+1}.$$

But this holds for all $j \geq 0$, so the corresponding stopping set $S(1)$ is empty. Since the optimal stopping set $S \subset S(1)$, it is also empty. Of course, a policy of always allowing the random process to continue never achieves a terminal reward, so it cannot be optimal. For any initial state i and any fixed integer $m > i$, we can achieve the reward r_m by waiting for the Markov chain to reach state m. However, no such policy is optimal since it can be improved by increasing the terminal state m.

Example 7.3 Now consider the same Markov chain as in the previous example and introduce a fixed transition cost $c_i = c > 0$. Let the rewards be given by

$$r_i = r\left\{1 - 2^{-i}\right\},$$

where r is a positive constant. In this case, the OSLA policy is to continue from state j if and only if

$$r_j < -c + \tfrac{1}{2}r_j + \tfrac{1}{2}r_{j+1},$$

which means that $r_{j+1} - r_j > 2c$. It is easily verified that this condition is equivalent to $2^j < r/4c$, and it follows that the stopping set

$$S(1) = \{m, m+1, m+2, \dots\},$$

where m is the integer determined by

$$m = \min \left\{ j \geq 0 : 2^j \geq r/4c \right\}. \tag{7.14}$$

For initial states $i < m$, the policy based on this stopping set amounts to waiting for the Markov chain to reach the state m. It is clear from (7.13) that the set $S(1)$ is closed, since transitions such as $m \to m - 1$ have zero probabilities. Hence, by Proposition 7.3, the OSLA rule is optimal.

We can also find the value function $\{v_i, i \geq 0\}$ for this example. For $i \geq m$, $v_i = r_i$ and for $0 \leq i < m$, we can evaluate v_i by calculating the expected number of transitions needed to reach m. Let $w_i = E\{T_i\}$, where T_i is the optimal stopping time. By conditioning on the first transition from i,

$$w_i = 1 + \tfrac{1}{2}w_i + \tfrac{1}{2}w_{i+1}, \qquad w_i = 2 + w_{i+1}.$$

Hence, $w_i = 4 + w_{i+2}$ and so on. Since $w_m = 0$, we obtain $w_i = 2(m - i)$. The expected net reward, starting at $i \leq m$ is given by $v_i = r_m - cw_i$,

$$v_i = r \left\{ 1 - \frac{1}{2^m} \right\} - 2c(m - i). \tag{7.15}$$

The value function is determined by this formula for $0 \leq i \leq m$ and

$$v_i = r \left\{ 1 - \frac{1}{2^i} \right\} \qquad (i \geq m).$$

The choice of the critical boundary state m can be approached in another way. It is not difficult to show that the integer m given by (7.14) maximizes the expression on the right of (7.15). Further calculation can be carried out to verify that we have determined a solution of the optimality equation

$$v_i = \max \left\{ r_i, -c + \tfrac{1}{2}v_i + \tfrac{1}{2}v_{i+1} \right\}$$

for each $i \geq 0$, if m is chosen according to (7.14).

Example 7.4 Imagine tossing a coin repeatedly and gambling on the accumulated difference between the numbers of Heads and Tails which occur. If there are r Heads after t tosses of the coin, the number of Tails must be $t - r$ and the difference is $j = 2r - t$. For this game of chance, the natural state variable is j and its possible values are $0, \pm 1, \pm 2 \cdots$. If the current state is j and the next toss yields a Head, then $j \to j + 1$. Similarly, if a Tail occurs, $j \to j - 1$. The possible transitions are

$$j \to j \pm 1, \text{ each with probability } \tfrac{1}{2}.$$

We are assuming that the coin is unbiased and the different tosses are independent of one-another. This process is a simple random walk on the positive and negative integers. It corresponds to the Markov chain with transition probabilities defined by

$$p_{j,j+1} = p_{j,j-1} = \tfrac{1}{2} \qquad (j = 0, \pm 1, \pm 2, \cdots).$$

Now suppose that there is a fixed cost $c > 0$ for each toss of the coin and that the terminal reward for stopping the game at the position j is $a|j|$. Then we have an optimal stopping problem for the simple random walk specified by

$$c_j = c, \qquad r_j = a|j|,$$

where a and c are positive constants. Here, the terminal rewards are not bounded above, so condition (7.7) is not satisfied. However, this is not a serious obstacle and the theory developed in Section 7.2 can be extended to allow for this.

A useful approach to this problem is to guess the form of the optimal stopping set and then modify its boundaries. The symmetry of the problem suggests that we consider stopping sets of the form

$$S = \{j : |j| \geq m\},$$

leaving the choice of the critical number $m \geq 0$ until later. Let $\{u_j\}$ be the expected net reward function for the policy based on S. We have $u_j = a|j|$ when $|j| \geq m$ and, otherwise,

$$u_j = am - cw_j. \tag{7.16}$$

For initial states j with $|j| < m$, termination must occur at $\pm m$ and the corresponding reward is am in either case. The expected cost of the transitions is cw_j, where $w_j = E\{T_j\}$ and T_j is the number of steps in the random walk from the initial state j until it stops at $\pm m$.

We must determine $w_j = E\{T_j\}$, for $|j| < m$. The technique of conditioning on the first step leads to the system of difference equations

$$w_j = 1 + \tfrac{1}{2}w_{j+1} + \tfrac{1}{2}w_{j-1},$$

and the boundary conditions are that $w_m = w_{-m} = 0$. By symmetry, $w_j = w_{-j}$ for $j = 1, 2, \ldots, m-1$. The solution of the difference equations, subject to these conditions, is given by

$$w_j = m^2 - j^2. \tag{7.17}$$

It is substantial exercise to prove this formula directly. However, it is easy to check that it gives a solution of the equations and that no other solution satisfies the boundary conditions: see Exercise 7.9.

For the stopping set S with boundaries at $\pm m$, the expected net rewards are given by (7.16) and (7.17). We obtain

$$u_j = am - cm^2 + cj^2. \qquad (7.18)$$

This holds in the continuation set where $|j| < m$ and also at the boundary points $j = \pm m$. In the stopping set, $|j| \geq m$ and

$$u_j = a|j|.$$

The expression on the right of (7.18) can be maximized, for fixed j, by choosing the critical number m. Consider the values of

$$g_m = am - cm^2 \qquad (m = 0, 1, 2, \dots)$$

and observe that

$$g_m - g_{m-1} = a + c - 2mc, \quad g_m - g_{m+1} = 2(m+1)c - (a+c).$$

A necessary condition that g_m is maximal is that $g_m \geq g_{m-1}$ and also that $g_m \geq g_{m+1}$. Hence, we must arrange that

$$m \leq \frac{(a+c)}{2c} \leq m+1. \qquad (7.19)$$

The critical number m is uniquely determined by these inequalities, except when the ratio $(a+c)/2c$ is a positive integer, in which case there are two possible choices for m. Notice that $m = 0$ if $a < c$ and $(a+c)/2c < 1$.

We have studied a restricted class of stopping sets S with a single parameter m. The policy determined by S is optimal within this class if m is chosen according to (7.19). This conclusion can now be strengthened because, in fact, we have obtained a solution of the optimality equation:

$$u_j = \max\left\{a|j|, -c + \tfrac{1}{2}u_{j+1} + \tfrac{1}{2}u_{j-1}\right\}. \qquad (7.20)$$

It needs further calculations in order to check this claim properly. For example, let $j = m$ and suppose that $m \geq 1$. The optimality equation requires that

$$u_m \;=\; am \geq -c + \tfrac{1}{2}u_{m+1} + \tfrac{1}{2}u_{m-1},$$

$$am \;\geq\; -c + \tfrac{1}{2}a(m+1) + \tfrac{1}{2}\left(am - cm^2 + c(m-1)^2\right).$$

The last inequality reduces to $m + 1 \geq (a+c)/2c$, which is true, according to (7.19).

Finally, it is worth noting that there are many other solutions of the optimality equation, but they do not represent attainable expectations. Let

$$h_j = b + cj^2 \qquad (j = 0, \pm 1, \pm 2, \cdots),$$

where b is a constant. The equation

$$h_j = -c + \tfrac{1}{2}h_{j+1} + \tfrac{1}{2}h_{j-1}$$

is satisfied, and it is clear that, if the constant b is large enough, $h_j \geq a|j|$ holds for all j, so we have a formal solution of (7.20).

EXERCISES

7.1 Verify the inequality (7.6) by using relation (7.5).

7.2 Show that the expected net reward attained by the OSLA rule is at least $u_i(1)$ for any initial state i.

7.3 Consider the value function $\{v_i, i \geq 0\}$ for Example 7.3. This is given by the formula (7.15) for $i \leq m$, where m is the integer determined by (7.14), and $v_i = r_i$ for $i \geq m$. Verify that the value function satisfies the optimality equation.

7.4 A Markov chain with states $0, 1, 2, \ldots$ has its transition probabilities defined by

$$p_{ii} = \frac{i}{i+1}, \qquad p_{i,i+1} = \frac{1}{i+1}$$

and all other $p_{ij} = 0$. Let $r_i = i$ and $c_i = 1/m$ for $i \geq 0$, where m is an integer, $m \geq 2$. Find the OSLA rule in this case and prove that it is optimal. Determine the value function.

7.5 A gambling game is defined in the following way. The player begins in state 0 and buys a number of moves. On reaching state $j = 0, \pm 1, \pm 2, \cdots$, he may stop and receive a reward of $6|j| - 9$ units or buy another move at cost 1. Each move $j \to j \pm 1$ is determined by tossing a coin. Find the optimal strategy and show that the game is fair in the sense that the net expectation of the player is zero if he uses it.

7.6 Consider a simple random walk on the integers $j = 0, \pm 1, \pm 2, \cdots$. Let $c > 0$ be the cost of a transition $j \to j \pm 1$, each having probability $\tfrac{1}{2}$. Let the reward for stopping the process at j be $(c + d)j^2$ where d is a positive constant. Find the expected net reward, starting at state 0, for the policy of waiting until the process first reaches one of the states $\pm m$. Deduce that there is no optimal policy.

7.7 Consider the simple random walk, as in the previous exercise. Suppose that there is a cost $s_j \geq 0$ for stopping the process at j and that the cost of the next transition is $c_j \geq 0$. Show that the minimum expected cost f_i, starting at i, must satisfy the optimality equation

$$f_i = \min\left\{s_i, c_i + \tfrac{1}{2}f_{i+1} + \tfrac{1}{2}f_{i-1}\right\}.$$

Define a sequence of approximations by

$$f_i(n) = \min\left\{s_i, c_i + \tfrac{1}{2}f_{i+1}(n-1) + \tfrac{1}{2}f_{i-1}(n-1)\right\}$$

for all i and $n = 1, 2, \ldots$. Let $f_i(0) = 0$ always. Show that this choice leads to an increasing sequence of approximations: $f_i(n+1) \geq f_i(n)$ always. How would you define $f_i(0)$ in order to obtain a decreasing sequence?

7.8 Verify that, for any real numbers α, β, γ,

$$\min(\alpha, \beta) - \min(\alpha, \gamma) \leq \max(0, \beta - \gamma).$$

Let $\{f_i\}$ and $\{g_i\}$ be two non-negative solutions of the optimality equation in Exercise 7.7. Define $h_i = f_i - g_i$ for $i = 0, \pm 1, \pm 2, \ldots$ and prove that

$$h_i \leq \max\left(0, \tfrac{1}{2}h_{i+1} + \tfrac{1}{2}h_{i-1}\right)$$

always holds. Now assume that the stopping cost $s_j \to 0$ as $j \to \infty$ and as $j \to -\infty$. By considering the states where h_i attains its maximum value, prove that $h_i \leq 0$ for all i. Deduce that $f_i = g_i$ always.

7.9 Let $w_j = 1 + \tfrac{1}{2}w_{j+1} + \tfrac{1}{2}w_{j-1}$ for $|j| < m$ and suppose that $w_m = w_{-m} = 0$. Show that the unique solution of these equations which satisfies the boundary conditions is given by (7.17): $w_j = m^2 - j^2$.

7.10 A long straight road has a parking place at each of the positions $1, 2, \ldots$. Each place has a known probability p $(0 < p < 1)$ of not being occupied, independently of the occupation of other positions. A driver, who can only drive in the direction of decreasing position numbers, will incur a cost n if he parks at position n and a cost $k > 1$ if he fails to park at all. As each place is reached, he cannot see whether or not places ahead of him are occupied. Find a policy for deciding where to park which minimizes the expected costs for the driver.

(i) Let v_i be the minimum expectation when the driver is at position i, but has not yet checked whether the parking place at i is free or not. Show that

$$v_i = (1 - p)\, v_{i-1} + p \min(i, v_{i-1})$$

for $i = 2, 3, \ldots$.

(ii) Deduce that the optimal policy is to park at position i if it is free and if $i \leq m$, for some integer $m \geq 1$.

(iii) Consider the case when $p = \tfrac{1}{2}$ and $k = 8$. Find the appropriate value of m.

8

Special Problems

8.1 INTRODUCTION

We turn now to some special problems of optimal stopping. The first is concerned with selling an asset. Imagine that you have a valuable asset for sale and that you receive a sequence of offers for it. How should you decide whether to accept a particular offer or to reject it and wait for the next? Several alternative approaches to this question will be described in the next section.

The problem described in Section 8.3 has been studied in many different versions. It is known variously as the best-choice problem, the marriage problem or the secretary problem. The aim is to find a decision procedure which maximizes the probability of selecting the best from a sequence of candidates. We shall consider mainly the standard form of the problem in which the number of candidates is known in advance.

Sequential decision problems are highly sensitive to the form and the timing of information given to the decision maker. The last section of this chapter is again concerned with selecting one from a sequence of rewards represented by random variables. Prophet inequalities are obtained by comparing the maximum expectation which can be obtained by a decision maker who observes the rewards sequentially with the maximum expected reward of a 'prophet' who can foresee all the alternative rewards before reaching a decision. We will examine the simplest of these inequalities and give a brief introduction to the subject, which is a growing area of research in sequential decision theory.

8.2 SELLING AN ASSET

Suppose that you have a valuable object for sale and that independent offers for it arrive at the times $1, 2, \ldots$. The offers are represented by random variables $X_1, X_2, \ldots$ and you may continue to receive them for as long as you wish before deciding to accept one of them. There is a cost $c > 0$ per unit time for waiting. At each stage, only the latest offer may be accepted and you must decide whether to accept it or to reject it and wait for the next one, at cost c.

We assume here that the independent offers are positive random variables with a known probability density function f. Let X be the latest one. The optimal decision rule for dealing with it does not depend explicitly on the time. We define $g(x)$, for $x \geq 0$, to be the maximum expected net gain that you can obtain, given that you have just received an offer $X = x$. If you accept it, the gain is x and no further offers are received. Otherwise, you must wait for the next offer Y and then maximize your expectation, given that $Y = y$, say. The principle of optimality shows that this maximum conditional expectation is $g(y)$, and it follows that

$$g(x) = \max \left\{ x, \int_0^\infty g(y) f(y)\, dy - c \right\}. \tag{8.1}$$

This is the optimality equation for the problem. Notice that the second term on the right does not depend on x: the constant

$$\xi = \int_0^\infty g(y) f(y)\, dy - c \tag{8.2}$$

represents the best that you can expect after rejecting all previous offers, up to and including x. Equation (8.1) becomes

$$g(x) = \max(x, \xi) \tag{8.3}$$

and this determines the optimal decision rule for the sequence of offers. A particular offer $X = x$ is worth accepting if $x \geq \xi$. Strictly speaking, both actions, accept and reject, are optimal if $x = \xi$, but this is immaterial since events such as $X = \xi$ have probability zero. For convenience, let us restrict attention to the optimal decision rule:

accept the first of the offers $X_1, X_2, \ldots, X_n$ such that $X_n \geq \xi$.

It remains to find the critical level ξ. As we shall see, there is a unique level $\xi > 0$ determined by (8.2) and (8.3), provided that the waiting cost c does not exceed the mean of the distribution of offers. From now on, it will be assumed that

$$0 < c < \mu = \int_0^\infty y f(y)\, dy \tag{8.4}$$

and that the mean μ is finite.

For any fixed $\xi > 0$, we can substitute $g(y) = \max(y, \xi)$ in the integal on the right of (8.2). Thus

$$\xi = \int_0^\xi \xi f(y)\, dy + \int_\xi^\infty y f(y)\, dy - c.$$

Since f is a probability density function,

$$\int_0^\infty f(y)\,dy = 1 \text{ and } \int_0^\xi f(y)\,dy = 1 - \int_\xi^\infty f(y)\,dy.$$

This fact can be used to simplify the expression for ξ and obtain the equation

$$\int_\xi^\infty (y - \xi) f(y)\,dy = c. \tag{8.5}$$

Let $L(\xi)$ denote the integral on the left here, regarded as a function of ξ. It can be shown that it has a derivative with respect to ξ, given by

$$L'(\xi) = -\int_\xi^\infty f(y)\,dy.$$

This is easy to verify under the assumption that the density $f(y)$ is continuous at $y = \xi$, but the extra assumption is not necessary. The derivative shows that $L(\xi)$ is strictly decreasing in ξ except where $\int_\xi^\infty f(y)\,dy = 0$, in which case $L(\xi) = 0$ also. Now

$$L(0) = \int_0^\infty yf(y)\,dy = \mu,$$

and we have shown that, as ξ increases through positive values, $L(\xi)$ is strictly decreasing for as long as it remains positive. It follows that equation (8.5), or equivalently $L(\xi) = c$, has a unique solution $\xi > 0$ if $0 < c < \mu$.

Condition (8.4) guarantees that it is worth waiting for the first offer. The expected net gain, starting at time zero, of accepting the offer X_1 after waiting until time 1 is $E\{X_1\} - c = \mu - c$. If this is positive, the optimal policy is to wait for an offer which exceeds the critical level ξ and this achieves the maximum expected net gain, given by

$$g(0) = \xi. \tag{8.6}$$

It is not difficult to show that $\xi \geq \mu - c$: see Exercise 8.1.

In practice, it is not easy to solve equation (8.5) explicitly for ξ, except in a few special cases.

Example 8.1 Suppose that each independent offer has the exponential distribution with probability density

$$f(x) = \lambda e^{-\lambda x} \qquad (x > 0).$$

The scale parameter $\lambda > 0$ is related to the mean $\mu = 1/\lambda$. In this case, equation (8.5) yields

$$\int_\xi^\infty (y - \xi) \lambda e^{-\lambda y}\,dy = c$$

and integration by parts reduces this to $\int_\xi^\infty e^{-\lambda y} dy = c$. The solution is given by

$$e^{-\lambda \xi} = \lambda c, \qquad\qquad \xi = -\lambda^{-1} \log (\lambda c).$$

This determines the critical level $\xi > 0$ provided that $\lambda c < 1$, which means that $c < 1/\lambda = \mu$.

Discounting future offers

The next example gives a slightly different formulation of the problem of selling an asset. Instead of using a fixed cost per unit time to allow for the effect of waiting, future rewards and their expectations will be discounted according to the delay. As before, there is a sequence of offers $X_1, X_2, \ldots$. In order to compare the latest offer $X_n = x$, say, with the next offer X_{n+1}, the value of the latter is discounted by a factor θ to allow for the delay. The discount factor θ is assumed to lie in the range $0 < \theta < 1$. If only one future offer is under consideration, we must compare x with $\theta E \{X_{n+1}\}$. The idea of discounting future rewards was introduced previously in Example 5.5, but here we also need to use expectations to assess the present value of unknown future rewards.

Example 8.2 A house is advertised for sale and the owner receives a sequence of offers $X_1, X_2, \ldots$, where the amounts are assumed to be independent positive random variables with a common probability density function f. We suppose further that the mean of this distribution is finite and that $f(x)$ is continuous for $x > 0$. Each offer must be accepted or rejected immediately after it is made and the sequence ends when an offer is accepted. The owner aims to maximize his discounted expectation from the sale, using a discount factor θ, $0 < \theta < 1$.

Let $g(x)$ be the discounted expectation from the sale, using an optimal policy, when he has just received the offer $X_n = x$ and all previous offers have been rejected. Either the amount x is accepted or it is rejected in order to wait for the next offer. In this case, if $X_{n+1} = y$, the next expectation at the time $n+1$ is $g(y)$ and this is effectively worth $\theta g(y)$ at time n, for comparison with x. This comparison leads to the optimality equation

$$g(x) = \max \left\{ x, \theta \int_0^\infty g(y) f(y) \, dy \right\}. \qquad (8.7)$$

Hence, $g(x) = \max(x, \xi)$ and the constant ξ satisfies

$$\xi = \theta \int_0^\infty g(y) f(y) \, dy. \qquad (8.8)$$

These equations can be treated in the same way as (8.1) and (8.2) in the first version of this problem. Since $g(y) = \xi$ for $0 \leq y \leq \xi$ and $g(y) = y$

otherwise, we have

$$\xi = \theta \left\{ \xi \int_0^\xi f(y)\, dy + \int_\xi^\infty y f(y)\, dy \right\},$$

and this can be rearranged by using the fact that $\int_0^\infty f(y)\, dy = 1$. Hence,

$$\xi = \theta\xi + \theta \int_\xi^\infty (y - \xi) f(y)\, dy. \tag{8.9}$$

It is a straightforward matter to check that this equation has a unique positive solution for the critical level ξ. Then, as before, the optimal policy is to wait for the first offer which exceeds ξ.

Equation (8.9) is equivalent to $R(\xi) = 0$, where

$$R(\xi) = \theta \int_\xi^\infty (y - \xi) f(y)\, dy - \xi(1 - \theta).$$

We note that $R(0) = \theta\mu$, where μ is the mean offer, so $R(0) > 0$. The function $R(\xi)$ is strictly decreasing in ξ for all $\xi > 0$, since its derivative is

$$R'(\xi) = -\theta \int_\xi^\infty f(y)\, dy - (1 - \theta) \le -(1 - \theta).$$

It is also clear that $R(\xi) \to -\infty$ as $\xi \to \infty$. Hence, there is a unique value of ξ such that $R(\xi) = 0$, as required.

Offers arriving in a Poisson process

So far, we have only considered random processes in discrete time. Here, it seems unrealistic to assume that the offers $X_1, X_2, \ldots$ arrive at regular intervals of time. There is a natural extension of the model in which the arrivals occur at random points of time. We suppose that the offers arrive in a Poisson process at an average rate $\lambda > 0$. This is the simplest form of random process in continuous time. A brief summary of the main properties is given below. For a more comprehensive treatment, see Grimmett and Stirzaker [GS82], for example. However, as we shall see, this extension of our model has little effect on the results we obtained earlier.

Consider selling an asset, assuming that offers are received at the time points of a Poisson process of rate λ. This means that the number of offers arriving in any interval $[a, b]$ is a Poisson random variable with mean $\lambda(b - a)$ and that the numbers received in non-overlapping intervals of time are independent of one another. At any time $t \ge 0$, future arrivals are independent of events in the past and the following property holds for short intervals of length δt:

(i)   the probability of a new offer arriving during $(t, t + \delta t)$ is $\lambda \delta t + o(\delta t)$, as $\delta t \to 0$.

Another property which will be used without proof is that waiting times for new arrivals always have an exponential distribution:

(ii)   at any time t, the random delay W before the arrival of the next offer has the probability density $\lambda e^{-\lambda w}$ for $w > 0$.

Hence, the mean waiting time is $E\{W\} = 1/\lambda$.

As before, it will be assumed that the successive offers have values $X_1, X_2, \ldots$ represented by independent positive random variables with a common probability density function f and finite mean $\mu = \int_0^\infty x f(x)\,dx$. The amounts offered are also assumed to be independent of the random process of arrival times. The cost of waiting is fixed at c per unit time. Each offer must be accepted or rejected immediately on arrival. In particular, rejection leads to an expected cost of waiting for the next offer given by

$$E\{cW\} = c/\lambda.$$

Let $g(x)$ be the maximum expected net gain, given that an offer $X = x$ has just been received and that all previous offers have been rejected. This net expectation allows for possible waiting costs in the future, but it does not include previous costs. Note that the decision points are the arrival times of successive offers generated by the Poisson process. The function $g(x)$ can be analysed by considering the immediate choice whether to stop and receive the amount x or to wait for the next offer Y, say. The maximum net expectation which can be achieved by waiting is $E\{g(Y)\} - c/\lambda$. Hence,

$$g(x) = \max\left\{x, \int_0^\infty g(y) f(y)\,dy - c/\lambda\right\}. \qquad (8.10)$$

This optimality equation is obtained in the same way as (8.1), except that the fixed cost c is now replaced by the expected cost of waiting for the next offer.

From now on, we can proceed exactly as before in the derivation of equations (8.2)–(8.5). Thus, $g(x) = \max(x, \xi)$, where

$$\xi = \int_0^\infty g(y) f(y)\,dy - c/\lambda.$$

The main conclusion is that the critical level ξ can be determined as the unique positive solution of the equation

$$\int_\xi^\infty (y - \xi) f(y)\,dy = c/\lambda, \qquad (8.11)$$

provided that $c < \lambda\mu$. This condition is equivalent to demanding that the expected net gain for waiting for and accepting the first offer is positive. Again, the optimal policy is to continue waiting until an offer is received which exceeds the level ξ.

Example 8.3 Suppose that the independent offers have a uniform distribution on the interval $[a, b]$, where $0 < a < b$. The probability density is given by

$$f(x) = (b - a)^{-1}$$

for $a \le x \le b$ and zero otherwise. The critical level ξ above which an offer is acceptable can be found by solving equation (8.11). Here, the notation can be simplified by rescaling the unit of time so that the arrival rate becomes $\lambda = 1$. In this case, equation (8.11) is precisely the same as equation (8.5), which applies when the offers are assumed to arrive at the fixed points $1, 2, 3, \ldots$ of discrete time. The optimal decision rule for the sequence of offers is not affected by whether the arrival times are treated as a Poisson process or as a regular succession.

When the cost parameter c is small enough, equation (8.11) yields values of ξ lying in the range $a < \xi < b$. In this case, the equation with $\lambda = 1$ becomes

$$\int_\xi^b (y - \xi)(b - a)^{-1} \, dy = c.$$

By evaluating the integral, we obtain

$$(b - \xi)^2 = 2(b - a)c$$

and the solution is

$$\xi = b - \{2(b - a)c\}^{\frac{1}{2}}. \qquad (8.12)$$

It is easy to see that the critical level ξ is a decreasing function of c which reduces from b to a as c increases through the interval $0 < c < (b - a)/2$.

When $c > (b - a)/2$, we find that $\xi < a$. In this case, equation (8.11) takes a different form:

$$\int_a^b (y - \xi)(b - a)^{-1} \, dy = c,$$

which is equivalent to $\mu - \xi = c$, where μ is the mean of the distribution of offers. Hence,

$$\xi = \mu - c, \qquad\qquad \mu = (a + b)/2. \qquad (8.13)$$

This result confirms that $\xi < a$ whenever $c > (b - a)/2$. Notice that offers $X_1, X_2, \ldots$ from the uniform distribution must all lie in the interval $[a, b]$ and, in particular, $X_1 \ge a > \xi$ occurs with probability 1. Any optimal decision procedure based on a critical level $\xi < a$ is equivalent to the rule:

wait for the first offer and accept it.

For this rule, the expected net reward is

$$E\{X_1\} - c = \mu - c.$$

Equations (8.6) and (8.13) show that

$$g(0) = \xi = \mu - c,$$

provided that $\xi \geq 0$. When $\mu < c$ in the formula (8.13), it is not optimal to wait for the first offer. Since $\xi < 0$, we have $g(0) = \max(0, \xi) = 0$. In this case, the cost of waiting is prohibitive and a better policy is to treat the asset as junk and throw it away.

8.3   THE MARRIAGE PROBLEM

Suppose that a woman receives several proposals of marriage over a period of time. Her problem is to decide, on each occasion, whether to accept the current proposal or to wait for the next. In a different scenario for the same problem, an interviewer is considering a sequence of candidates for a vacant secretarial post and the aim is to find the best candidate for the job. This has received a good deal of attention in the research literature: a review by Freeman [Fre83] describes the historical background and many different extensions of the problem. We shall be mainly concerned with the original form and its solution.

In the standard form of the marriage or secretary problem, n candidates are arranged in random order and interviewed one at a time. It is assumed that the interviewer knows the relative ranking of present and previous candidates at each stage and she must decide immediately whether to accept or reject the present candidate. No recall of previous candidates is permitted and the interviews continue until one of the candidates is accepted or all of them have been rejected. The aim is to maximize the probability of choosing the best of the n candidates.

Here n is given in advance and we suppose that $n \geq 2$. The assumption that the candidates are interviewed in random order is crucial: all the $n!$ possible orderings are equally likely. In effect, we have an optimal stopping problem for a Markov chain with $2n - 1$ states. Consider the situation when r candidates have already been interviewed, $r = 1, 2, \ldots, n$. For $r = 1$, the first candidate must be the best so far. If $r \geq 2$, it is important to distinguish two different states according to whether the rth candidate is observed to be the best so far, or not. Someone who turns out to be best so far will be called a leading candidate. A person who is not a leading candidate at stage r cannot possibly be the best of all n, so there is no point is deciding to stop and accept such a candidate. For any $r < n$, it may be preferable to reject the present candidate, even in the case of a leading candidate. If we wait for the next candidate, the conditional probability of finding that the $(r+1)$th is a leading candidate is just $1/(r+1)$. It is not difficult to verify that this holds for any $r = 1, 2, \ldots, n-1$, as a consequence of the random ordering of candidates.

Let us define f_r to be the maximum of the conditional probability of eventually selecting the best of all the candidates, given that the rth is a leading candidate and that previous ones have been rejected. Similarly, g_r is the corresponding maximum probability of choosing the best candidate, given that the one considered at stage r is not the best so far. Thus, f_r and g_r are the conditional probabilities of success in selecting the best candidate, using an optimal policy from stage r onwards, $r = 1, 2, \ldots, n$. We note that $f_n = 1$ and $g_n = 0$, since success is assured in one situation and failure in the other. The main aim is to determine f_1 and the corresponding optimal selection procedure. Notice that g_1 does not represent a proper state of the Markov process since the first candidate is always leading, but it will be useful to keep the notation.

In general, g_r represents the situation when r candidates have been rejected and then there is a transition to a new state represented by f_{r+1} or g_{r+1}. The transition probabilities are $1/(r+1)$ and $r/(r+1)$, respectively, since the chance of obtaining a leading candidate next is $1/(r+1)$. It follows that

$$g_r = \frac{1}{r+1} f_{r+1} + \frac{r}{(r+1)} g_{r+1}, \qquad (8.14)$$

and this holds for $r = 1, 2, \ldots, n-1$, with the boundary conditions $f_n = 1, g_n = 0$.

The optimality equation for the problem can be specified by considering f_r. The immediate decision is whether or not to accept the present candidate, who is known to be the best of the first r. The probability of obtaining the best of all n at this stage is r/n, since this is the probability that the best candidate of all is also the leading one among the first r. On the other hand, rejection means that $f_r = g_r$ and the probability of eventual success then satisfies equation (8.14). The optimal decision corresponds to

$$f_r = \max \left\{ \frac{r}{n}, g_r \right\}, \qquad (8.15)$$

which shows that there is a definite advantage in waiting for the next candidate if $g_r > r/n$. The optimality equation is valid for $r = 1, 2, \ldots, n$.

In turns out that the optimal decision procedure for this problem has a special form which makes it much easier to solve the equations (8.14) and (8.15). The special form can be deduced from the following result.

Proposition 8.1 *If $f_r = r/n$ for some $r < n$, then $f_{r+1} = (r+1)/n$.*

Proof This will be established by proving that $f_r > r/n$ holds whenever $f_{r+1} > (r+1)/n$. The stated proposition follows immediately by using equation (8.15).

Suppose that $f_{r+1} > (r+1)/n$. Equation (8.15), with r replaced by $r+1$, shows that $f_{r+1} = g_{r+1}$ and then (8.14) yields $g_r = g_{r+1}$. We now have $g_r = f_{r+1} > (r+1)/n$. It follows that $g_r > r/n$ and hence, by (8,15), that $f_r = g_r > r/n$. This completes the proof. $\square$

We can now define the critical number s by

$$s = \min\{r \geq 1 : f_r = r/n\}.$$

It is clear that $1 \leq s \leq n$, since $f_n = 1$. Equation (8.15) and the definition of s mean that

$$f_r = g_r > \frac{r}{n} \qquad\qquad (r = 1, 2, \ldots s - 1) \qquad\qquad (8.16)$$

and the proposition shows that

$$f_r = \frac{r}{n} \geq g_r \qquad\qquad (r = s, s+1, \ldots, n). \qquad\qquad (8.17)$$

It also follows from (8.15) that an optimal selection procedure is determined by the rule:

> reject the first $s-1$ candidates and accept the first leading candidate among those interviewed at stages $s, s+1, \ldots, n$.

This procedure fails if no leading candidate is found after the initial rejections. In this case, the nth candidate cannot be the best, but must be accepted.

We are now in a position to determine s.

Proposition 8.2 *For any $n \geq 2$, the critical number s is given by*

$$s = \min\left\{r \geq 1 : \frac{1}{n-1} + \frac{1}{n-2} + \cdots + \frac{1}{r} \leq 1\right\}.$$

Proof It follows from equation (8.14) that $g_r = g_{r+1}$ whenever $f_{r+1} = g_{r+1}$, so we can deduce from (8.16) that

$$f_1 = g_1 = f_2 = g_2 = \cdots = f_{s-1} = g_{s-1}. \qquad\qquad (8.18)$$

Then equations (8.14) and (8.17) yield

$$g_r = \frac{1}{n} + \frac{r}{r+1} g_{r+1} \qquad\qquad (s-1 \leq r \leq n-1). \qquad\qquad (8.19)$$

These equations can be solved for $r = n-1, n-2$, and so on by using the fact that $g_n = 0$. Thus $g_{n-1} = 1/n$,

$$g_{n-2} = \frac{1}{n} + \frac{n-2}{n-1} g_{n-1} = \frac{n-2}{n}\left\{\frac{1}{n-2} + \frac{1}{n-1}\right\}.$$

It is easily verified by backwards induction that

$$g_r = \frac{r}{n}\left\{\frac{1}{r} + \frac{1}{r+1} + \cdots + \frac{1}{n-1}\right\} \tag{8.20}$$

for $s - 1 \leq r \leq n - 1$. Finally, we can use two optimality conditions obtained from the relations (8.16) and (8.17):

$$g_{s-1} > \frac{s-1}{n}, \qquad g_s \leq \frac{s}{n}.$$

According to (8.20), these conditions are equivalent to

$$\frac{1}{s-1} + \frac{1}{s} + \cdots + \frac{1}{n-1} > 1 \geq \frac{1}{s} + \frac{1}{s+1} + \cdots + \frac{1}{n-1}. \tag{8.21}$$

This confirms the definition of s in the proposition. $\square$

Equations (8.18) and (8.20) obtained in the course of this proof also determine f_1, the probability of discovering the best of all n candidates by applying the optimal selection rule. This is given by

$$f_1 = g_{s-1} = \frac{s-1}{n}\left\{\frac{1}{s-1} + \frac{1}{s} + \cdots + \frac{1}{n-1}\right\}. \tag{8.22}$$

The inequalities (8.21) and equation (8.22) can be used to determine the critical number s and the maximum probability of success f_1 for any particular $n \geq 2$. When n is large, the sums of reciprocals which occur in the formulae can be replaced by integrals. For example,

$$\frac{1}{s} + \frac{1}{s+1} + \cdots + \frac{1}{n-1}$$

is well approximated by

$$\int_s^n \frac{1}{y}\,dy = \log\left(\frac{n}{s}\right).$$

Hence, the ratio n/s must be close to $e = 2.718\ldots$. It is not difficult to show that both s/n and f_1 converge to $1/e = 0.368\ldots$ as $n \to \infty$: see Exercise 8.7. The table below shows the values of s and f_1 obtained for specified values of n. Notice that the probability of selecting the best candidate never falls below the limit $1/e$.

n	s	f_1
2	1	0.500
3	2	0.500
4	2	0.458
5	3	0.433
6	3	0.428
7	3	0.414
8	4	0.410
9	4	0.406
10	4	0.399
15	6	0.389
20	8	0.384
30	12	0.379
50	19	0.374

Extensions of the problem

There are many different variations and extensions of the problem. For example, the aim of minimizing the expected rank of the selected candidate has been investigated. Another variation is based on relaxing the assumption that an offer made to any candidate must be accepted. These and other extensions are described in the review paper by Freeman [Fre83]. One basic assumption in the standard form of the marriage problem is that the number of candidates is known in advance. It has been shown that, if n is unknown, the range of admissible policies is very large. We shall conclude this section with a brief explanation of this result.

Let us imagine a decision maker attempting to choose the best of a sequence of n candidates, where n is an unknown positive integer. We retain the assumption that, conditional on the value of n, all the $n!$ possible arrangements of the candidates are equally likely. Clearly, there is no point in choosing someone who is not a leading candidate, so we shall restrict attention to selection rules based on leading candidates. Consider a general policy

$$q = \{q_1, q_2, \dots\},$$

with $0 \le q_r \le 1$ for $r \ge 1$, which is applied as follows: choose the rth candidate with conditional probability q_r if all the previous ones have been rejected and if the r th turns out to be the best so far; otherwise consider the next candidate if one is available. It is important to note that this selection process fails automatically if at any stage a candidate is rejected and then it is found that there are no more to come.

For a policy q and any $r \ge 1$, let

$$u_r(q) = (1 - q_1)\left(1 - \frac{q_2}{2}\right)\cdots\left(1 - \frac{q_r}{r}\right). \tag{8.23}$$

This is the probability of rejecting all of the first r candidates when $n \ge r$. The formula is easy to verify by induction. Since q_1 is the probability of accepting the first candidate, $u_1(q) = (1 - q_1)$. Now suppose that (8.23) is valid for some $r \geq 1$ and also that $n \geq r+1$. After rejecting r candidates, the $(r + 1)$th is a leading candidate with probability $1/(r + 1)$ and the probability of this event together with a decision to accept the candidate is $q_{r+1}/(r + 1)$. Hence,

$$u_{r+1}(q) = u_r(q)\left(1 - \frac{q_{r+1}}{r + 1}\right)$$

and this is enough to complete the induction.

We define $v_n(q)$, for each possible $n \geq 1$, as the probability of success associated with the policy when the number of candidates is n. It can be shown that

$$v_n(q) = \frac{1}{n}\sum_{r=1}^{n} u_{r-1}(q)\, q_r,$$

where $u_o(q) = 1$, but this formula will not be needed. The policy q is said to be *admissible* if there is no other policy q' with values

$$v_n(q') \geq v_n(q)$$

for all $n \geq 1$ and with the strict inequality holding in at least one case. The main result established in a paper by Abdel-Hamid, Bather and Trustrum [AHBT82] is that a policy q is admissible if and only if

$$u_r(q) \to 0 \qquad \text{as} \qquad r \to \infty.$$

This condition has a natural interpretation. Given that there are n candidates, $u_n(q)$ is the probability of rejecting all of them. In general, we have

$$1 \geq u_1(q) \geq u_2(q) \geq \cdots$$

and the limit is an infinite product $\lambda = \lambda(q)$. If $\lambda > 0$, the policy has an obvious weakness in that $u_n(q)$, the probability of failure by default, is at least λ for any value of n. This weakness is removed if $\lambda = 0$, but it may seem surprising that such a mild condition is enough to guarantee that the policy is admissible.

8.4 PROPHET INEQUALITIES

Suppose that you are offered the choice of one from a set of n possible prizes. Each prize has a value represented by a random variable. The aim here is to compare three different ways of regulating the information available to you before you reach a decision. Let us assume that the random variables $X_1, X_2, \ldots, X_n$, representing the values, are non-negative and independent of one another. Their probability distributions are given, so we know the expected values in advance. Let

$$\mu_j = E\{X_j\} \qquad (1 \le j \le n).$$

Each μ_j is positive and we suppose that all these means are finite.

First consider the situation when you are allowed to inspect all the prizes before making your choice. In this case, it is clear that the maximum possible expectation is

$$u_n = E\{\max(X_1, X_2, \ldots, X_n)\}.$$

Now suppose that you must make up your mind without inspecting any of the prizes. Then the best that you can do is to find the maximum of $\mu_1, \mu_2, \ldots, \mu_n$ and the expected reward is

$$w_n = \max(\mu_1, \mu_2, \ldots, \mu_n).$$

Finally, consider the sequential case : suppose that you may inspect the prizes one at a time but, at each stage, you must reject those examined so far before you are allowed to see the next. For convenience, let us imagine that the values of $X_n, X_{n-1}, \ldots, X_1$ are revealed sequentially, in that order, until you decide to stop and choose the one you have just observed. Let v_n be the maximum expectation, attainable by using an optimal stopping rule. Similarly, for any $r \le n$, v_r is the maximum expectation which can be achieved when there are r prizes left to consider, represented by $X_r, X_{r-1}, \ldots, X_1$. We can evaluate $v_1, v_2, \ldots$ by dynamic programming. Thus

$$v_1 = \mu_1, \qquad v_2 = E\{\max(X_2, v_1)\}.$$

The second follows by considering whether to accept X_2 or reject it and wait to see X_1. In general,

$$v_r = E\{\max(X_r, v_{r-1})\}, \tag{8.24}$$

for $r = 2, 3, \ldots, n$.

It is intuitively obvious that

$$u_n \ge v_n \ge w_n, \tag{8.25}$$

with equality when $n = 1$. As we shall see, u_n and v_n can be much larger than w_n. The growth of u_n and v_n as n increases will be illustrated by examples where all the underlying random variables have the same distribution. Then we shall investigate how much of an advantage can be obtained from having full information, rather than sequential information. The main result is that

$$u_n \le 2v_n \tag{8.26}$$

always holds.

The relation (8.26) is a prophet inequality: it shows that a sequential decision maker can always expect to gain at least half as much as a prophet who is able to foresee the future, and make a choice with full information on all the values available. There are many results of this type, but we shall restrict attention to just one. A survey of prophet inequalities can be found in the paper by Hill and Kertz [HK92]. Proofs of the basic inequalities (8.25) and (8.26) will be given later, after the examples.

Any random variable X can be described by its distribution function F, which is defined by

$$F(x) = P(X \le x)$$

for all real x. For non-negative random variables, $F(x) = 0$ when $x < 0$ and, if X has a probability density function f,

$$F(x) = \int_0^x f(y)\, dy$$

when $x \ge 0$. The mean can be expressed in the form

$$E\{X\} = \int_0^\infty (1 - F(x))\, dx. \tag{8.27}$$

We need to establish a generalization of this formula which will be useful in what follows. For any constant $v \ge 0$,

$$E\{\max(X, v)\} = v + \int_v^\infty (1 - F(x))\, dx. \tag{8.28}$$

This is valid for any non-negative random variable with a finite mean, but it will be assumed that X has a probability density f, in order to simplify the proof. We have

$$E\{\max(X, v)\} = \int_0^v vf(x)\, dx + \int_v^\infty xf(x)\, dx.$$

and, by using the fact that $\int_0^\infty f(x)\, dx = 1$,

$$E\{\max(X, v)\} = v + \int_v^\infty (x - v) f(x)\, dx.$$

We note that the integral here must converge, since it is bounded above by the mean $\mu = \int_0^\infty xf(x)\, dx$. Now write $(x - v) = \int_v^x 1\, dy$ and consider the double integral obtained, which is over the ranges $v < y < x$ and $x > v$. The order of integration can be reversed and this leads to

$$\int_v^\infty (x - v) f(x)\, dx = \int_v^\infty \left[\int_y^\infty f(x)\, dx \right] dy.$$

Since $\int_y^\infty f(x)\,dx = 1 - F(y)$, we obtain

$$E\{\max(X,v)\} = v + \int_v^\infty (1 - F(y))\,dy,$$

as required.

The examples will illustrate the behaviour of the sequences $\{u_n\}$ and $\{v_n\}$ when the independent random variables $X_1, X_2, \ldots, X_n$ all have the same distribution, with distribution function F. Note that the distribution function of $\max(X_1, X_2, \ldots, X_n)$ can be found by using the product law for independent events.

$$\begin{aligned} P(\max(X_1, X_2, \ldots, X_n) \le x) &= P(X_j \le x, 1 \le j \le n) \\ &= P(X_1 \le x)\,P(X_2 \le x)\ldots P(X_n \le x) \\ &= F^n(x). \end{aligned}$$

Hence, by using (8.27),

$$u_n = E\{\max(X_1, X_2, \ldots, X_n)\} = \int_0^\infty (1 - F^n(x))\,dx. \qquad (8.29)$$

Example 8.4 Suppose that $X_1, X_2, \ldots$ are independent and uniformly distributed on the interval $[0,1]$. Their common probability density is given by $f(x) = 1$ for $0 \le x \le 1$ and $f(x) = 0$ otherwise. Hence,

$$F(x) = x \qquad\qquad (0 \le x \le 1),$$

but $F(x) = 0$ for $x < 0$ and $F(x) = 1$ for $x > 1$. It follows from equation (8.29) that

$$u_n = \int_0^1 (1 - x^n)\,dx = \frac{n}{n+1}.$$

The common mean of all the random variables is $\mu = \frac{1}{2}$, so $w_n = \frac{1}{2}$ for all $n \ge 1$. The sequence $\{v_n\}$ starts with $v_1 = \frac{1}{2}$ and then we can use the relations (8.24) and (8.28) to obtain

$$v_n = v_{n-1} + \int_{v_{n-1}}^1 (1 - x)\,dx = \frac{1}{2}\left(1 + v_{n-1}^2\right),$$

for $n \ge 2$. The table below gives some values of u_n and v_n obtained from these formulae.

n	1	2	3	4	5	10	20
u_n	0.500	0.667	0.750	0.800	0.833	0.909	0.952
v_n	0.500	0.625	0.695	0.742	0.775	0.861	0.920

It can be shown that $v_n \geq (n+1)/(n+3)$ always holds: see Exercise 8.8. In other words,

$$u_n = 1 - \frac{1}{n+1}, \qquad v_n \geq 1 - \frac{2}{n+3}.$$

Example 8.5 Now consider sampling values from an exponential distribution which has no upper bound. Let

$$f(x) = e^{-x}, \qquad F(x) = 1 - e^{-x} \qquad (x \geq 0).$$

Here, the mean $\mu = 1$, so $w_n = 1$ for all $n \geq 1$. The formula (8.29) shows that

$$u_n = \int_0^\infty \left\{ 1 - \left(1 - e^{-x}\right)^n \right\} dx$$

and this can be evaluated by using the substitution $y = 1 - e^{-x}$. We find that

$$u_n = 1 + \frac{1}{2} + \frac{1}{3} + \cdots + \frac{1}{n}.$$

The recurrence relation for the sequential gains is easily obtained from (8.28):

$$v_n = v_{n-1} + e^{-v_{n-1}}$$

for $n \geq 2$ and $v_1 = u_1 = 1$. In this example, both u_n and v_n are of order $\log n$ for large n. The sequence $\{u_n\}$ is well known: in particular, the difference $u_n - \log n$ converges to a limit $\gamma = 0.577\ldots$ known as Euler's constant. Let $a = e - 1 = 1.718\ldots$. Then the following inequalities are valid for $n \geq 1$:

$$\log(a + n) \leq v_n \leq u_n \leq 1 + \log n.$$

The last of these is an easy exercise, but the first is more difficult.

We now consider a simple discrete distribution which shows that u_n and v_n can be of order n. Note that, since

$$\max(X_1, X_2, \ldots, X_n) \leq X_1 + X_2 + \cdots + X_n,$$

the corresponding expectations must satisfy

$$u_n \leq \mu_1 + \mu_2 + \cdots + \mu_n$$

and so $u_n \leq n\mu$ when each $\mu_j = \mu$.

Example 8.6 Let K be a large positive constant and suppose that each of the random variables $X_1, X_2, \ldots$ takes the values 0 and K with probabilities $1 - K^{-1}$ and K^{-1}, respectively. Then the common mean $\mu = 1$. In general,

$w_n = 1$ and we also know that $u_n \leq n$. Here, $\max{(X_1, X_2, \ldots, X_n)}$ has only two possible values, 0 and K. Hence

$$u_n = K\{1 - P(X_1 = X_2 = \cdots = X_n = 0)\},$$
$$u_n = K\left\{1 - \left(1 - K^{-1}\right)^n\right\}.$$

It is also easy to see that $v_n = u_n$ for each $n \geq 1$. This can be established by using the recurrence relation (8.24). We have $v_1 = 1$, and it is clear that v_{r-1} always lies between 0 and K for $r \geq 2$, so

$$v_r = 1 + \left(1 - K^{-1}\right) v_{r-1}.$$

It is a straightforward matter to show by induction that

$$v_n = u_n = K\left\{1 - \left(1 - K^{-1}\right)^n\right\}.$$

For any fixed $n \geq 1$, this expression can be made as close as we please to n, by choosing a sufficiently large value for the constant K.

We now return to the basic inequalities (8.25) and (8.26). They will be established for any non-negative random variables $X_1, X_2, \ldots$ with finite means, but possibly different distributions.

Proposition 8.3 *Let $X_1, X_2, \ldots$ be independent, non-negative random variables with finite means $\mu_1, \mu_2, \ldots$. For each $n \geq 1$, let*

$$u_n = E\{\max X_1, X_2, \ldots, X_n\}, \qquad w_n = \max(\mu_1, \mu_2, \ldots, \mu_n)$$

and define the sequence $\{v_n\}$ by

$$v_1 = \mu_1, \qquad v_n = E\{\max(X_n, v_{n-1})\} \qquad\qquad (n \geq 2).$$

Then the following inequalities hold:

$$u_n \geq v_n \geq w_n$$

for all n, with equality when $n = 1$.

Proof It is easy to verify that $v_n \geq w_n$. We have $v_1 = w_1 = \mu_1$ and then $v_2 = E\{\max(X_2, v_1)\}$, so $v_2 \geq E\{X_2\} = \mu_2$ and also $v_2 \geq v_1$. Similarly, we obtain $v_n \geq \mu_n$ and $v_n \geq v_{n-1}$ for all $n \geq 2$. Hence, $v_1 \leq v_2 \leq \cdots \leq v_n$ and finally,

$$v_n = \max(v_1, v_2, \ldots, v_n) \geq \max(\mu_1, \mu_2, \ldots, \mu_n) = w_n.$$

It is more awkward to prove that $u_n \geq v_n$ always. We shall need to rely on a fundamental property of conditional expectations. For any random variables X and Y, if $E\{Y\}$ is finite, it can be evaluated by conditioning on X:

$$E\{Y\} = E\{h(X)\}, \quad \text{where } h(x) = E\{Y \mid X = x\}$$

for each possible value $X = x$.

Fix $n \geq 2$ and define $X = X_n$, $Y = \max(X_1, X_2, \ldots, X_n)$. Then we have $u_n = E\{Y\} = E\{h(X)\}$ and

$$h(x) = E\{Y \mid X_n = x\} = E\{\max(X_1, X_2, \ldots, X_{n-1}, x)\}.$$

Since $X_1, X_2, \ldots, X_{n-1}$ are independent of X_n, there is no need for conditioning in the last expectation and it follows that

$$h(x) \geq E\{\max(X_1, X_2, \ldots, X_{n-1})\} = u_{n-1}.$$

Clearly $h(x) \geq x$ always holds, so we obtain

$$u_n = E\{h(X)\} \geq E\{\max(X_n, u_{n-1})\}.$$

Successive applications of this inequality lead to the required result:

$$u_2 \geq E\{\max(X_2, u_1)\} = v_2,$$

since $u_1 = v_1$. Then

$$u_3 \geq E\{\max(X_3, u_2)\} \geq E\{\max(X_3, v_2)\} = v_3$$

and so on. $\qquad\qquad\qquad\qquad\qquad\qquad\qquad\qquad\qquad\qquad\quad\square$

Proposition 8.4 *Let $X_1, X_2, \ldots$ be non-negative random variables with finite means $\mu_1, \mu_2, \ldots$ and let u_n and v_n be defined as before, for $n \geq 1$. Then*

$$u_n \leq v_{n-1} + v_n \leq 2v_n.$$

Proof Let $Y_1 = X_1$ and $Y_r = \max(X_r, v_{r-1})$ for $r = 2, 3, \ldots, n$. We also define $v_0 = 0$ and

$$Z_r = \max(X_r - v_{r-1}, 0)$$

so that $Y_r = v_{r-1} + Z_r$. We can use the fact that $v_0 \leq v_1 \leq \cdots \leq v_{n-1}$, established in the proof of the previous proposition, to obtain

$$Y_r \leq v_{n-1} + Z_r \qquad (1 \leq r \leq n).$$

It is clear from the definitions that $Y_r \geq X_r$ always. It follows that

$$\max (X_1, X_2, \ldots, X_n) \leq \max (Y_1, Y_2, \ldots, Y_n)$$
$$\leq v_{n-1} + \max (Z_1, Z_2, \ldots, Z_n),$$

and $u_n = E\{\max (X_1, X_2, \ldots, X_n)\}$ satisfies

$$u_n \leq v_{n-1} + E\{\max (Z_1, Z_2, \ldots, Z_n)\}.$$

Every $Z_r \geq 0$, so

$$\max (Z_1, Z_2, \ldots, Z_n) \leq Z_1 + Z_2 + \cdots + Z_n.$$

However, $Z_r = Y_r - v_{r-1}$ and $E\{Z_r\} = v_r - v_{r-1}$ since $v_r = E\{\max (X_r, v_{r-1})\} = E\{Y_r\}$, by definition. Hence

$$E\{Z_1 + Z_2 + \cdots + Z_n\} = v_1 - v_0 + v_2 - v_1 + \cdots + v_n - v_{n-1}$$

and this sum reduces to v_n. Finally, by using the previous inequalities, we find that

$$u_n \leq v_{n-1} + E\{Z_1 + Z_2 + \cdots + Z_n\} = v_{n-1} + v_n$$

and the proof is complete. $\square$

Previously, we have assumed that $X_1, X_2, \ldots$ are independent random variables, but the above proposition holds more generally. However, the independence assumption is needed for the interpretation of v_n as the maximum expected gain that can be obtained from choosing one of the values $X_n, X_{n-1}, \ldots, X_1$, when they are revealed sequentially.

EXERCISES

8.1 Show that the critical level ξ, determined by equation (8.5) when $\mu > c$, always satisfies $\xi \geq \mu - c$.

8.2 Consider the problem of selling an asset when $c = £10$ and the probability density of the independent offers is

$$f(x) = a^{-1} \exp\{-(x - b)a^{-1}\} \qquad (x \geq b)$$

and zero for $x < b$, $a = £200$ and $b = £1000$. Find the optimal policy in this case.

8.3 Solve equation (8.5) for the critical level ξ in terms of c, when $0 < c < 1$ and

$$f(y) = 2(1 + y)^{-3} \qquad (y \geq 0).$$

8.4 In Example 8.2, suppose that the probability density is $f(x) = \lambda e^{-\lambda x}$ for $x \geq 0$, where λ is a positive constant. Show that the critical value ξ satisfies the equation

$$\lambda \xi e^{\lambda \xi} = \theta/(1-\theta).$$

8.5 In the problem of selling an asset, suppose that recall of previous offers is permitted at any stage. Consider a situation when n offers $x_1, x_2, \ldots, x_n$ are available and set $y = \max(x_1, x_2, \ldots, x_n)$. Show that the OSLA rule is to accept the maximum offer y if and only if $y \geq \xi$, where the critical level ξ is determined by equation (8.5), as before. In fact, this is the optimal policy, but this is harder to prove. In other words, allowing recall makes no real difference to the problem.

8.6 In the marriage problem verify the formula (8.20) by using (8.19) and the fact that $g_n = 0$.

8.7 Show that $\int_{k-1}^{k} \frac{1}{t}\, dt$ lies between $1/k$ and $1/(k-1)$ for any integer $k \geq 2$. Hence, prove that

$$\log\left(\frac{n}{s}\right) < \frac{1}{s} + \frac{1}{s+1} + \cdots + \frac{1}{n-1}$$
$$< \frac{1}{s-1} + \frac{1}{s} + \cdots + \frac{1}{n-1} < \log\left(\frac{n-1}{s-2}\right).$$

Deduce that the critical number $s = s(n)$ for the marriage problem satisfies $s/n \to 1/e$ as $n \to \infty$.

8.8 Prove that $v_n \geq (n+1)/(n+3)$ in Example 8.4.

8.9 In Example 8.5, show that

$$\log(a+n) \leq v_n \leq u_n \leq 1 + \log n$$

for all $n \geq 1$, where $a = e - 1 = 1.718\ldots$.

Part III

Markov Decision Processes

9

General Theory

9.1 INTRODUCTION

The application of dynamic programming to stochastic processes was discussed in Chapter 6 and the Markov system described in Section 6.2 covers a large class of models. However, Part II of this book is mainly restricted to optimal stopping problems, where the choice at each stage is whether to stop a random process or let it continue. Some of the most interesting and useful applications of sequential decision theory involve controlling the motion of the underlying process in a more flexible way, by choosing a sequence of actions from a set which includes many possibilities.

The first investigations of Markov decision processes were stimulated by Howard's book [How60], and there is now a substantial literature. The theory is concerned with Markov chains in discrete time, where there is a choice of alternative transition matrices. The aim is to minimize expected costs over an infinite future and there are two alternative ways of dealing with this. The first is to assume that future costs are discounted, which ensures that the total future expectation is always finite. The technique of backwards induction works well here. The basic results for the discounted case will be established in this chapter. The second approach to minimizing future expectations is to investigate the long-term average costs, without discounting. It is possible to treat this as a limit of the discounted case, but complications occur and a careful analysis is required. The minimization of average costs will be studied in Chapter 10.

Since we are dealing with Markov chains, matrix notation is useful and it is also helpful to represent future expectations by column vectors, with each component giving the conditional expectation for a particular initial state. It will be assumed that the state space is finite, so that transition matrices and expectation vectors are finite dimensional. Many of the main results can be extended to infinitely many states without much difficulty, at least when future costs are discounted. Section 9.2 begins by introducing the special notation and the idea of minimizing the different components of a vector simultaneously. Successive iterations of this technique, for finite horizons, lead to a unique limit vector which satisfies the optimality equation for an infinite future horizon. There is another way of solving the optimality equation which relies on a policy improvement algorithm. This will be described and illustrated in Section 9.3. The final section considers a replacement problem based on a generalization of the simple model used in Chapter 4. Here, the structure of the problem suggests a simple form of policy, and the optimal policy can be found by verifying that it corresponds to a solution of the optimality equation.

9.2   MINIMIZING DISCOUNTED EXPECTATIONS

We shall be concerned with the following general problem. Consider a process in discrete time with a finite set of possible states, $S = \{1, 2, \ldots, s\}$. From any state $j \in S$, the transition to the next state is controlled by choosing a row vector

$$\mathbf{p}_j = (p_{j1}, p_{j2}, \ldots, p_{js})$$

from a prescribed set D_j of probability distributions on S. Any selection $\mathbf{p}_j \in D_j$, for each $j = 1, 2, \ldots, s$, determines the rows of a stochastic matrix $P = (p_{jk})$, which means that

$$p_{jk} \geq 0 \qquad \text{and} \qquad \sum_{k=1}^{s} p_{jk} = 1$$

always. In other words, every row of P is a probability distribution on the state space S. We suppose that each set D_j is finite, which yields a finite set $D = D_1 \times D_2 \times \cdots \times D_s$ of possible transition matrices $P \in D$. The main point here is that the choice of a transition matrix P, for the next transition, is always restricted to a finite set and that the rows are chosen separately. The cost of a single transition depends on $P \in D$. For each state $j \in S$ and any choice of $\mathbf{p}_j \in D_j$, the immediate cost

$$c_j = c_j (p_{j1}, p_{j2}, \ldots, p_{js})$$

is prescribed. This determines a column vector $\mathbf{c} = (c_1, c_2, \ldots, c_s)'$ associated with each policy P. The aim is to control the Markov system by choosing a sequence of policies to minimize the expected future costs over a period. We shall rely on discounting to ensure that the total expectation remains finite when this period becomes infinite. Let θ be the discount factor and suppose that $\theta < 1$.

The simplest type of control procedure is to use a stationary policy P in which the transition matrix P remains constant. In this case, every transition is controlled by the appropriate row of P, depending on the current state but not explicitly on the time. Let $\mathbf{c}$ be the cost vector associated with P

and consider the corresponding stationary policy. The total discounted future expectation is represented by the vector

$$\mathbf{v} = \mathbf{c} + \theta P \mathbf{c} + \theta^2 P^2 \mathbf{c} + \cdots \tag{9.1}$$

In order to see this, let i be the initial state of the process. The immediate cost incurred for the first transition is c_i and this leads to a new state j with probability p_{ij}. The expectation of the discounted cost for the second transition is $\theta \sum_{j=1}^{s} p_{ij} c_j$. Similarly, for $t \geq 2$, the distribution of the state after t transitions is determined by the ith row of the matrix P^t, as we saw in Chapter 7. Hence, the discounted cost of the $(t+1)$th transition is $\theta^t c_k$, if the state at time t is k. Conditional on the initial state i at time 0, the expectation is given by summing over all possible $k \in S$, obtaining $\theta^t \sum p_{ik}^{(t)} c_k$, where $p_{ik}^{(t)}$ is the appropriate element of the product matrix P^t. It follows that the total discounted expectation over an infinite future is just the ith component of the vector

$$\mathbf{v} = \sum_{t=0}^{\infty} \theta^t P^t \mathbf{c}.$$

The set of possible transition costs c_j is finite and bounded, since the decision sets D_j and the state space are finite. We are only interested in the relative costs of different decisions and such comparisons are not affected by adding the same constant to all of them. This device can be used, if necessary, to ensure that all the costs are non-negative. We may assume without loss of generality, that all the prescribed transition costs satisfy

$$0 \leq c_j \leq b, \tag{9.2}$$

where b is a positive constant.

In order to demonstrate the use of vector inequalities, let us verify that every component of the total discounted expectation in (9.1) satisfies

$$0 \leq v_j \leq \frac{b}{1-\theta}.$$

For any two $s \times 1$ column vectors $\mathbf{v}$ and $\mathbf{u}$, we say that

$$\mathbf{v} \leq \mathbf{u} \quad \text{if} \quad v_j \leq u_j \quad \text{for} \quad j = 1, 2, \ldots s,$$

but not otherwise. Here, we need to show that

$$\mathbf{v} = \mathbf{c} + \theta P \mathbf{c} + \theta^2 P^2 \mathbf{c} + \cdots \leq (1-\theta)^{-1} b \mathbf{1},$$

where $\mathbf{1}$ is the unit vector with s components. Similarly, let $\mathbf{0}$ be the column vector of s zeros. By assumption (9.2), we have $\mathbf{0} \leq \mathbf{c} \leq b\mathbf{1}$. Recall that P is a stochastic matrix with non-negative elements and row sums equal to 1. Thus, $P\mathbf{1} = \mathbf{1}$ and hence $P^2\mathbf{1} = P(P\mathbf{1}) = \mathbf{1}$ and so on. It is easy to see that $\mathbf{c} \leq b\mathbf{1}$ leads to the inequality

$$P\mathbf{c} \leq P(b\mathbf{1}) = bP\mathbf{1} = b\mathbf{1}.$$

The equivalent relations for the ith component are

$$\sum_j p_{ij}c_j \leq \sum_j p_{ij}b = b\sum_j p_{ij} = b.$$

All these sums are over the state space, $j = 1, 2, \ldots, s$. Notice that the inequalities rely on the fact that the terms are non-negative. Returning to vector notation, we can use $P\mathbf{c} \leq b\mathbf{1}$ to obtain $P^2\mathbf{c} \leq P(b\mathbf{1}) = b\mathbf{1}$. In general, $P^t\mathbf{c} \leq b\mathbf{1}$ for $t \geq 1$, and this is also true for $t = 0$ if we interpret $P^0 = I$, the identity matrix. It follows that

$$\mathbf{v} = \sum_{t=0}^{\infty} \theta^t P^t \mathbf{c} \leq \sum_{t=0}^{\infty} \theta^t b\mathbf{1}$$

and, by summing the geometric series, that

$$\mathbf{v} \leq (1 - \theta)^{-1} b\mathbf{1}.$$

This shows that the infinite series of vectors used for the total discounted expectation in (9.1) must always converge. Every term of the series is a non-negative vector, so $\mathbf{v} \geq \mathbf{0}$, and the partial sums form a non-decreasing sequence of vectors.

The problem is to find a sequence of policies, or preferably a stationary policy, to minimize the total expected discounted cost. Our first approach will be to consider the minimum expectation over a future period of length n and then let $n \to \infty$. Apart from the special notation here, this is just another application of backwards induction.

Successive approximations
In dynamic programming, we begin by defining a function which represents the minimum expectation over n transitions in terms of the initial state $i \in S$. Here, it is convenient to treat this minimum future expectation as the i th component of a vector $\mathbf{x}(n)$. We set $\mathbf{x}(0) = \mathbf{0}$ and consider the vectors $\mathbf{x}(1), \mathbf{x}(2) \ldots$, determined by successive applications of the principle of optimality. For $n \geq 1$, we have

$$\mathbf{x}(n) = \min\{\mathbf{c} + \theta P\mathbf{x}(n - 1)\}. \tag{9.3}$$

It is important to understand this minimization operation before proceeding further. Each of the components $x_i(n)$ is obtained from the vector $\mathbf{x}(n - 1)$

by minimizing with respect to the choice of $\mathbf{p}_i \in D_i$, which also fixes the cost c_i for the next transition from state i. Thus,

$$x_i(n) = \min_{D_i} \left\{ c_i + \theta \sum_j p_{ij} x_j(n-1) \right\} \qquad (1 \le i \le s).$$

The value of the discounted expectation on the right is determined from the components of $\mathbf{x}(n-1)$ by the cost c_i and the probability distribution $\mathbf{p}_i = (p_{i1}, p_{i2}, \dots, p_{is})$ associated with a particular choice from D_i. An optimal choice, which attains the minimum, is always possible since D_i is finite. The operation implicit in relation (9.3) evaluates the components of $\mathbf{x}(n)$ by separate minimizations. This usually involves substantial computations: a very simple worked example is given below. However, the main point is that the sequence of vectors $\{\mathbf{x}(n)\}$ is well defined. As we shall see later, this sequence converges to a limit vector $\mathbf{u}$ which is the solution of the optimality equation

$$\mathbf{u} = \min\{\mathbf{c} + \theta P \mathbf{u}\}. \tag{9.4}$$

Example 9.1 Let $S = \{1, 2, 3\}$ and let the discount factor $\theta = \frac{1}{2}$. For each state i, there are two possible decisions. The corresponding transition probability vectors $\mathbf{p}_i$ and costs c_i are shown below.

State	Decision			Cost
1	$\frac{3}{4}$	$\frac{1}{4}$	0	16
	$\frac{1}{2}$	0	$\frac{1}{2}$	16
2	0	$\frac{1}{2}$	$\frac{1}{2}$	7
	$\frac{1}{2}$	$\frac{1}{2}$	0	4
3	$\frac{1}{2}$	$\frac{1}{2}$	0	0
	0	$\frac{1}{2}$	$\frac{1}{2}$	4

The first few expected cost vectors can be found without much difficulty, but it is worth checking the calculations to see the application of relation (9.3) in detail. We have $\mathbf{x}(0) = \mathbf{0}$ and $\mathbf{x}(1) = \min\{\mathbf{c}\}$, which means that each component $x_i(1)$ is obtained by minimizing the immediate cost c_i. Hence,

$$\mathbf{x}(1) = (16, 4, 0)'.$$

The evaluation of $\mathbf{x}(2)$ needs a more careful comparison of the alternative expectations:

$$x_1(2) = \min\left\{16 + \frac{1}{2}(12+1), \quad 16 + \frac{1}{2}(8+0)\right\},$$

$$x_2(2) = \min\left\{7 + \frac{1}{2}(2+0), \quad 4 + \frac{1}{2}(8+2)\right\},$$

$$x_3(2) = \min\left\{0 + \frac{1}{2}(8+2), \quad 4 + \frac{1}{2}(2+0)\right\}.$$

For $x_1(2)$, the second of the two alternative decisions yields the minimum; for $x_2(2)$, it is the first; for $x_3(2)$, either choice will do. Hence

$$\mathbf{x}(2) = (20, 8, 5)'.$$

Similarly, for $\mathbf{x}(3)$, we have

$$x_1(3) = \min\left\{16 + \frac{1}{2}(15+2), \quad 16 + \frac{1}{2}(10+2.5)\right\},$$

$$x_2(3) = \min\left\{7 + \frac{1}{2}(4+2.5), \quad 4 + \frac{1}{2}(10+4)\right\},$$

$$x_3(3) = \min\left\{0 + \frac{1}{2}(10+4), \quad 4 + \frac{1}{2}(4+2.5)\right\}.$$

Here, the minimum is given by the second alternative for $x_1(3)$ and by the first choice for $x_2(3)$ and $x_3(3)$. We find that

$$\mathbf{x}(3) = (22.25, \ 10.25, \ 7.0)'.$$

The sequence of vectors $\{\mathbf{x}(n)\}$ increases to a limit $\mathbf{u}$, whose components u_i give the minimum future expectation for particular initial states i. It will be shown later by a different method that, in this case,

$$\mathbf{u} = (24.4, \ 12.4, \ 9.2)'.$$

We now return to the general problem and continue our investigation of the successive approximations $\{\mathbf{x}(n)\}$ generated by relation (9.3). Notice that the minimization operation used to obtain $\mathbf{x}(n)$ from $\mathbf{x}(n-1)$ also determines a pair, consisting of a policy P and its cost vector $\mathbf{c}$, which attain the minimum on the right of (9.3). The minimizing policy may not be unique and it usually depends on n. We also note that, since $\mathbf{x}(n)$ represents the minimum discounted expectation over n transitions, it cannot exceed the corresponding expectation for a stationary policy represented by a fixed pair $(P, \mathbf{c})$ over the same period. We have already shown that these expectations have an upper bound, and it follows that

$$\mathbf{x}(n) \le \mathbf{c} + \theta P\mathbf{c} + \cdots + \theta^{n-1}P^{n-1}\mathbf{c} \le (1-\theta)^{-1}b\mathbf{1}$$

for all $n \geq 1$. The proposition below shows that $\mathbf{x}(n) \rightarrow \mathbf{u}$ as $n \rightarrow \infty$, where the limit vector $\mathbf{u}$ satisfies the optimality equation for an infinite future horizon. The next result establishes the uniqueness of this solution and the final proposition shows how the solution can be used to find a stationary policy which is optimal in the long run.

Proposition 9.1 *The sequence of vectors defined by relation* (9.3) *is non-decreasing:*

$$0 \leq \mathbf{x}(1) \leq \mathbf{x}(2) \leq \cdots$$

and $\mathbf{x}(n) \rightarrow \mathbf{u}$ *as* $n \rightarrow \infty$, *where*

$$\mathbf{u} = \min\{\mathbf{c} + \theta P\mathbf{u}\}.$$

Proof Since $\mathbf{x}(0) = \mathbf{0}$, $\mathbf{x}(1) = \min\{\mathbf{c}\}$ and, in view of the assumption (9.2), $\mathbf{x}(1) \geq \mathbf{0}$. Now assume the inductive hypothesis that $\mathbf{x}(n) \geq \mathbf{x}(n-1)$ for some $n \geq 1$ and consider the components of $\mathbf{x}(n+1)$. We have

$$x_i(n+1) = \min_{D_i}\left\{c_i + \theta \sum_j p_{ij}x_j(n)\right\}.$$

Since θ and all the transition probabilities are non-negative,

$$c_i + \theta \sum_j p_{ij}x_j(n) \geq c_i + \theta \sum_j p_{ij}x_j(n-1)$$

for any choice of $\mathbf{p}_i \in D_i$. Hence,

$$x_i(n+1) \geq \min_{D_i}\left\{c_i + \theta \sum_j p_{ij}x_j(n-1)\right\} = x_i(n)$$

and this holds for each state i, which means that $\mathbf{x}(n+1) \geq \mathbf{x}(n)$, as required. Each sequence $\{x_i(n), n = 1, 2, \dots\}$ is non-decreasing and bounded above by $b/(1-\theta)$. Let us define

$$u_i = \lim_{n \to \infty} x_i(n) \qquad\qquad (1 \leq i \leq s).$$

Then $\mathbf{x}(n) \rightarrow \mathbf{u} = (u_1, u_2, \dots, u_n)'$ as $n \rightarrow \infty$.

The asymptotic behaviour is complicated by the fact that the policies which attain the minimum in (9.3) may depend on n. For any fixed policy P and the corresponding cost vector $\mathbf{c}$, we have

$$\mathbf{x}(n) \leq \mathbf{c} + \theta P\mathbf{x}(n-1)$$

and, by letting $n \to \infty$, we obtain

$$\mathbf{u} \leq \mathbf{c} + \theta P \mathbf{u}.$$

Since this holds for any admissible pair $(P, \mathbf{c})$ it follows that

$$\mathbf{u} \leq \min \left\{ \mathbf{c} + \theta P \mathbf{u} \right\}.$$

In order to complete the proof, it will be enough to show that the reverse inequality also holds. We know that

$$u_i \geq x_i(n) = \min_{D_i} \left\{ c_i + \theta \sum_j p_{ij} x_j(n-1) \right\}$$

always holds. For any fixed $\varepsilon > 0$, since $x_j(n-1) \to u_j$ as $n \to \infty$, and since the state space is finite, we know that $x_j(n-1) \geq u_j - \varepsilon$ for all $j \varepsilon S$ when n is sufficiently large. Suppose this is true for some integer $n = n(\varepsilon)$. Then

$$c_i + \theta \sum_j p_{ij} x_j(n-1) \geq c_i + \theta \sum_j p_{ij} u_j - \theta \varepsilon.$$

This holds for any choice of $\mathbf{p}_i \varepsilon D_i$, so we obtain

$$u_i \geq x_i(n) \geq \min_{D_i} \left\{ c_i + \theta \sum_j p_{ij} u_j \right\} - \theta \varepsilon$$

and, since $0 < \theta < 1$.

$$u_i \geq \min_{D_i} \left\{ c_i + \theta \sum_j p_{ij} u_j \right\} - \varepsilon.$$

The last inequality is valid for each $i \varepsilon S$ and any $\varepsilon > 0$, which implies that

$$\mathbf{u} \geq \min \left\{ \mathbf{c} + \theta P \mathbf{u} \right\}.$$

This completes the proof that $\mathbf{u}$ is a solution of the optimality equation. □

Proposition 9.2 *The limit vector $\mathbf{u}$ defined in Proposition 9.1 is the unique solution of the optimality equation:*

$$\mathbf{u} = \min \left\{ \mathbf{c} + \theta P \mathbf{u} \right\}.$$

Proof Let $\mathbf{v}$ be any other vector such that

$$\mathbf{v} = \min \left\{ \mathbf{c} + \theta P \mathbf{v} \right\}.$$

Define $M = \max_{i \varepsilon S} |v_i - u_i|$. Then $M \geq 0$ and $M = 0$ only if $\mathbf{v} = \mathbf{u}$. Suppose that the maximum is attained when $i = m \varepsilon S$. We may assume, without loss of generality, that $v_m - u_m = M$. Otherwise, the vectors $\mathbf{v}$ and $\mathbf{u}$ can be interchanged. Now let $Q = (q_{ij})$ be the optimal transition matrix associated with $\mathbf{u}$ and let $\mathbf{d}$ be the corresponding cost vector. In other words,

$$\mathbf{u} = \mathbf{d} + \theta Q \mathbf{u} \leq \mathbf{c} + \theta P \mathbf{u}$$

and the inequality holds for any other admissible pair $(P, \mathbf{c})$. For the other solution $\mathbf{v}$, the pair $(Q, \mathbf{d})$ need not be optimal, so we have

$$\mathbf{v} \leq \mathbf{d} + \theta Q \mathbf{v}.$$

It follows that

$$\mathbf{v} - \mathbf{u} \leq \{\mathbf{d} + \theta Q \mathbf{v}\} - \{\mathbf{d} + \theta Q \mathbf{u}\} = \theta Q (\mathbf{v} - \mathbf{u})$$

In particular, the mth component of the difference satisfies

$$v_m - u_m \leq \theta \sum_j q_{mj} (v_j - u_j).$$

But Q is a stochastic matrix and $v_j - u_j \leq M$, for all $j \in S$, by definition of M. Hence

$$M = v_m - u_m \leq \theta M,$$

which means that $M (1 - \theta) \leq 0$ and, since $\theta < 1$, the only possible conclusion is that $M = 0$. We have proved that $\mathbf{v} = \mathbf{u}$, so the solution of the optimality equation is unique.   □

We are now in a position to show that the optimality equation (9.4) determines a stationary policy which is optimal in the sense that its performance cannot be improved by any other admissible decision procedure. Consider a procedure defined by a policy sequence $\{P^{(t)}\}$ and the corresponding vectors $\{\mathbf{c}^{(t)}\}$, where $\mathbf{c}^{(t)}$ represents the cost of choosing the transition matrix $P^{(t)}$ for the tth transition, $t = 1, 2, \ldots$. For this procedure, the probability of reaching state j at time t from state i at time 0 is given by the (i, j)th element of the product matrix $P^{(1)} P^{(2)} \ldots P^{(t)}$. The total discounted expected cost over n transitions, conditional on the initial state i, is the ith component of the vector

$$\mathbf{y}(n) = \mathbf{c}^{(1)} + \theta P^{(1)} \mathbf{c}^{(2)} + \cdots + \theta^{n-1} P^{(1)} P^{(2)} \ldots P^{(n-1)} \mathbf{c}^{(n)}$$

and the sequence of vectors $\{\mathbf{y}(n)\}$ converges to the limit

$$\mathbf{y} = \mathbf{c}^{(1)} + \sum_{t=1}^{\infty} \theta^t P^{(1)} P^{(2)} \ldots P^{(t)} \mathbf{c}^{(t+1)}. \tag{9.5}$$

This infinite series is a generalization of the formula (9.1), which gives the total discounted expectation for a stationary policy. Its convergence can be established in the same way, by using the fact that it is a sum of non-negative vectors, and the upper bounds provided by assumption (9.2).

Proposition 9.3 *Let* **u** *be the solution of the optimality equation and suppose that the minimum is attained by the pair* $(Q, \mathbf{d})$, *where* $Q \in D$ *and* **d** *is the corresponding cost vector. The transition matrix* Q *determines a stationary policy and its total discounted expectation over an infinite period is represented by* **u**. *Let* $\{P^{(t)}\}$ *be a policy sequence with* $P^{(t)} \in D$ *for* $t \geq 1$ *and let* **y** *be the corresponding total expectation vector defined by equation* (9.5). *Then* $\mathbf{u} \leq \mathbf{y}$, *for any such policy sequence, and hence the stationary policy* Q *is optimal.*

Proof We have $\mathbf{u} = \mathbf{d} + \theta Q \mathbf{u}$ and repeated application of this equation leads to

$$\mathbf{u} = \mathbf{d} + \theta Q\mathbf{d} + \cdots + \theta^{n-1} Q^{n-1} \mathbf{d} + \theta^n Q^n \mathbf{u}.$$

The vector **u** is the limit of the sequence $\{\mathbf{x}(n)\}$ and we showed earlier that $0 \leq \mathbf{x}(n) \leq (1 - \theta)^{-1} b\mathbf{1}$ always. Hence, **u** satisfies the same conditions and, since Q^n is a stochastic matrix for each $n \geq 1$, the components of $Q^n \mathbf{u}$ must always lie between 0 and $(1 - \theta)^{-1} b$. It follows that $\theta^n Q^n \mathbf{u} \to \mathbf{0}$ as $n \to \infty$ and

$$\mathbf{u} = \mathbf{d} + \theta Q\mathbf{d} + \theta^2 Q^2 \mathbf{d} + \cdots,$$

which is the total discounted expectation for the stationary policy Q, according to (9.1).

For each $n, \mathbf{y}(n)$ is the total discounted expected cost obtained over n transitions by using the policy sequence $\{P^{(t)}\}$. On the other hand, our construction of the sequence $\{\mathbf{x}(n)\}$ based on relation (9.3), ensures that each $\mathbf{x}(n)$ is the minimum discounted expectation over n transition. Hence $\mathbf{x}(n) \leq \mathbf{y}(n)$, for $n \geq 1$. Since $\mathbf{x}(n) \to \mathbf{u}$ and $\mathbf{y}(n) \to \mathbf{y}$ as $n \to \infty$, it follows that $\mathbf{u} \leq \mathbf{y}$. $\qquad \square$

9.3   POLICY IMPROVEMENTS

Proposition 9.3 shows that we can restrict attention to stationary policies in our search for an optimal control procedure. The method of successive approximations produces a sequence of vectors converging to the solution of the optimality equation. However, there is another approach which leads more directly to the optimal policy in a finite number of iterations. This is due to Howard [How60], and it is based on repeating a technique which replaces any given policy by a better one, until an optimal policy is achieved.

We need a preliminary result which shows how to evaluate a given policy. For a stationary policy determined by the pair $(P, \mathbf{c})$, the total discounted expectation vector $\mathbf{v}$ is defined by the formula (9.1). This convergent series can be rearranged to obtain

$$\mathbf{v} = \mathbf{c} + \theta P \left[\mathbf{c} + \theta P \mathbf{c} + \theta^2 P^2 \mathbf{c} + \cdots \right]$$

In other words,

$$\mathbf{v} = \mathbf{c} + \theta P \mathbf{v}. \tag{9.6}$$

Proposition 9.4 *Let P be a stochastic matrix and let θ be a scalar such that $0 < \theta < 1$. Then the matrix $I - \theta P$ is non-singular and its inverse can be expressed in the form:*

$$[I - \theta P]^{-1} = I + \theta P + \theta^2 P^2 + \cdots .$$

Proof Let $\mathbf{x}$ be a column vector such that

$$[I - \theta P] \mathbf{x} = \mathbf{0}.$$

In order to verify that the matrix here is non-singular, we must show that $\mathbf{x} = \mathbf{0}$ is the only possible solution. The equation means that $\mathbf{x} = \theta P \mathbf{x}$ and this relation can be used repeatedly:

$$\mathbf{x} = \theta \mathbf{P} \mathbf{x} = \theta P \left[\theta P \mathbf{x} \right] = \theta^2 P^2 \mathbf{x}$$

and so on. It is easy to see that

$$\mathbf{x} = \theta^n P^n \mathbf{x},$$

for all $n \geq 1$. But each element of the matrix P^n is a transition probability and hence the elements of $\theta^n P^n$ cannot exceed θ^n. This implies that $\theta^n P^n$ must converge to the zero matrix as $n \to \infty$. It follows that

$$\mathbf{x} = \lim_{n \to \infty} \theta^n P^n \mathbf{x} = \mathbf{0}.$$

Now consider the matrix identity:

$$[I - \theta P] \left[I + \theta P + \cdots + \theta^{n-1} P^{n-1} \right] = I - \theta^n P^n.$$

This holds for every $n \geq 1$ and, when $n \to \infty$, we obtain

$$[I - \theta P] \left[I + \theta P + \theta^2 P^2 + \cdots \right] = I$$

The series of matrices on the left must converge to a limit and the equation shows that the limit matrix is $[I - \theta P]^{-1}$. $\qquad \square$

This result shows that equation (9.6) can be solved for the vector $\mathbf{v}$ by inverting a matrix. We have $[I - \theta P]\,\mathbf{v} = \mathbf{c}$ and

$$\mathbf{v} = [I - \theta P]^{-1}\,\mathbf{c}. \tag{9.7}$$

It will be explained how to construct a sequence of stationary policies $\{P^{(r)}, r = 1, 2, \dots\}$. Each $P^{(r)} \in D$ and it has a prescribed vector $\mathbf{c}^{(r)}$ of immediate costs and a vector $\mathbf{v}^{(r)}$ of total discounted expectations, which can be found by substituting $P^{(r)}$ and $\mathbf{c}^{(r)}$ in the formula (9.7).

We begin with an arbitrary policy $P^{(1)} \in D$ and the corresponding vectors $\mathbf{c}^{(1)}$ and $\mathbf{v}^{(1)}$. If this policy is suboptimal, it can be replaced by a new policy $P^{(2)} \in D$ which is an improvement in the sense that $\mathbf{v}^{(2)} \le \mathbf{v}^{(1)}$. The new policy is selected by examining the quantity

$$\min_{D_i} \left\{ c_i + \theta \sum_j p_{ij} v_j^{(1)} \right\}$$

for each $i \in S$. We choose the row $\mathbf{p}_i = (p_{i1}, p_{i2}, \dots p_{is})$ of the policy matrix $P^{(2)}$ by minimizing the total expected cost incurred by a single transition according to $\mathbf{p}_i$, followed by infinitely many future transitions using the old policy $P^{(1)}$. The discounted future expectation after the first transition from state i is described by the components of the vector $\mathbf{v}^{(1)}$. As we shall see, the effect of improving the first transition in this way is to produce a better stationary policy for the whole future. Since the vector $\mathbf{v}^{(1)}$ satisfies (9.6),

$$\mathbf{v}_i^{(1)} = c_i^{(1)} + \theta \sum_j p_{ij}^{(1)} v_j^{(1)}$$

and it follows that

$$\min_{D_i} \left\{ c_i + \theta \sum_j p_{ij} v_j^{(1)} \right\} \le v_i^{(1)} \tag{9.8}$$

for $i = 1, 2, \dots, s$. If the strict inequality applies, it indicates that a definite improvement can be achieved and the i th row of $P^{(2)}$ is defined by choosing any $\mathbf{p}_i \in D_i$ which attains the minimum. However, if equality holds in (9.8), the i th row of $P^{(1)}$ attains the minimum and there is no advantage in changing it. In fact, it is helpful to adopt a convention that the ith row should not be changed in such cases.

Policy iteration

The policy improvement technique can be summarized in the following way:

Let $P^{(1)} \in D$ be a stationary policy and let $\mathbf{c}^{(1)}$ and $\mathbf{v}^{(1)}$ be the corresponding cost vectors. For each $i \in S$ if

$$\min_{D_i} \left\{ c_i + \theta \sum_j p_{ij} v_j^{(1)} \right\} < v_i^{(1)}, \qquad (9.9)$$

replace the ith row of $P^{(1)}$ by using any $\mathbf{p}_i \in D_i$ which achieves the minimum as the ith row of $P^{(2)}$. Otherwise, retain the ith row of $P^{(1)}$ in $P^{(2)}$.

One immediate consequence of the convention is that there is no change in the policy $P^{(1)}$ unless the strict inequality (9.9) holds for some i. Hence, if $P^{(2)} = P^{(1)}$, we can infer that

$$\mathbf{v}^{(1)} = \min \left\{ \mathbf{c} + \theta P \mathbf{v}^{(1)} \right\}$$

and conclude that the policy $P^{(1)}$ is optimal. Otherwise, $P^{(2)} \neq P^{(1)}$ and the policy iteration must lead to some improvement in performance.

Proposition 9.5 *Let $P^{(2)}$ be the new policy obtained from $P^{(1)}$ by policy iteration. Let $\mathbf{c}^{(r)}$ and $\mathbf{v}^{(r)}$ be the cost vectors associated with $P^{(r)}, r = 1, 2$. Then $\mathbf{v}^{(2)} \leq \mathbf{v}^{(1)}$, so the new policy is an improvement on $P^{(1)}$. Further, $v_i^{(2)} < v_i^{(1)}$ for some i, unless $P^{(2)} = P^{(1)}$.*

Proof The inequalities (9.8) show that

$$\mathbf{v}^{(1)} \geq \min \left\{ \mathbf{c} + \theta P \mathbf{v}^{(1)} \right\}$$

and, since the new policy $P^{(2)}$ is chosen to achieve the minimum on the right,

$$\mathbf{v}^{(1)} \geq \mathbf{c}^{(2)} + \theta P^{(2)} \mathbf{v}^{(1)}.$$

This relation can be used repeatedly to obtain a succession of inequalities:

$$\mathbf{c}^{(2)} + \theta P^{(2)} \mathbf{v}^{(1)} \geq \mathbf{c}^{(2)} + \theta P^{(2)} \mathbf{c}^{(2)} + \theta^2 \left(P^{(2)} \right)^2 \mathbf{v}^{(1)}$$

$$\geq \mathbf{c}^{(2)} + \theta P^{(2)} \mathbf{c}^{(2)} + \cdots + \theta^{n-1} \left(P^{(2)} \right)^{n-1} \mathbf{c}^{(2)}$$

$$+ \theta^n \left(P^{(2)} \right)^n \mathbf{v}^{(1)}$$

for each $n \geq 3$. The last term here converges to the zero vector as $n \to \infty$, since every component is of order θ^n. It follows that in the limit we obtain

$$\mathbf{c}^{(2)} + \theta P^{(2)} \mathbf{v}^{(1)} \geq \mathbf{c}^{(2)} + \theta P^{(2)} \mathbf{c}^{(2)} + \theta^2 \left(P^{(2)} \right)^2 \mathbf{c}^{(2)} + \cdots$$

and, according to (9.1), the sum of this series is $\mathbf{v}^{(2)}$. We have shown that policy iteration leads to the inequalities

$$\mathbf{v}^{(1)} \geq \mathbf{c}^{(2)} + \theta P^{(2)} \mathbf{v}^{(1)} \geq \mathbf{v}^{(2)}$$

as required. Finally, suppose that $P^{(2)} \neq P^{(1)}$ and let the two transition matrices differ in row i, say. Then the policy iteration procedure guarantees that

$$v_i^{(1)} > c_i^{(2)} + \theta \sum_j p_{ij}^{(2)} v_j^{(1)} \geq v_i^{(2)}$$

and hence $v_i^{(1)} > v_i^{(2)}$.   □

Now consider a sequence of stationary policies $\{P^{(r)}, r = 1, 2, \ldots\}$ constructed by successive policy improvements, starting with an arbitrary policy $P^{(1)} \in D$. Let $\mathbf{c}^{(r)}$ and $\mathbf{v}^{(r)}$ be the cost vectors associated with $P^{(r)}$, where

$$\mathbf{v}^{(r)} = \left[I - \theta P^{(r)} \right]^{-1} \mathbf{c}^{(r)} \tag{9.10}$$

gives the total discounted expectation for any initial state. Proposition 9.5 shows that each policy iteration produces an improvement in performance:

$$\mathbf{v}^{(1)} \geq \mathbf{v}^{(2)} \geq \mathbf{v}^{(3)} \geq \cdots$$

But we know that the set D of admissible policies is finite, so the sequence $\{\mathbf{v}^{(r)}\}$ can only contain a finite number of distinct vectors. Since the sequence is decreasing, we must have $\mathbf{v}^{(r+1)} = \mathbf{v}^{(r)}$ for some r. Define

$$k = \min \left\{ r \geq 1 : \mathbf{v}^{(r+1)} = \mathbf{v}^{(r)} \right\}.$$

Then the vectors $\mathbf{v}^{(1)} \geq \mathbf{v}^{(2)} \geq \cdots \geq \mathbf{v}^{(k)}$ are distinct, and it follows that the policies $P^{(1)}, P^{(2)}, \ldots P^{(k)}$ are all different. Then we have $\mathbf{v}^{(k+1)} = \mathbf{v}^{(k)}$ and $P^{(k+1)} = P^{(k)}$: after several policy improvements, no further improvement is possible and the final policy is optimal. The proof of Proposition 9.5 shows that, for any policy iteration

$$\mathbf{v}^{(r+1)} \leq \min \left\{ \mathbf{c} + \theta P \mathbf{v}^{(r)} \right\} \leq \mathbf{v}^{(r)}.$$

Since $\mathbf{v}^{(k+1)} = \mathbf{v}^{(k)}$, it is a solution of the optimality equation:

$$\mathbf{v}^{(k)} = \min \left\{ \mathbf{c} + \theta P \mathbf{v}^{(k)} \right\}.$$

We also established earlier, in Propositions 9.2 and 9.3, that the solution is unique and that a policy which attains the minimum is optimal. Hence, $P^{(k)}$ is an optimal stationary policy.

We now return to the example described in the previous section, in order to illustrate the method of policy improvements. Further examples and more interesting applications can be found in Howard's book [How60].

Example 9.2 The specification is the same as in Example 9.1. There are three states and two possible decisions in each of them. The transition probabilities and costs are repeated here.

State	Decision			Cost
1	$\frac{3}{4}$	$\frac{1}{4}$	0	16
	$\frac{1}{2}$	0	$\frac{1}{2}$	16
2	0	$\frac{1}{2}$	$\frac{1}{2}$	7
	$\frac{1}{2}$	$\frac{1}{2}$	0	4
3	$\frac{1}{2}$	$\frac{1}{2}$	0	0
	0	$\frac{1}{2}$	$\frac{1}{2}$	4

As before, let $\theta = \frac{1}{2}$. We begin with the policy formed by choosing the first alternative decision in each of the states:

$$P^{(1)} = \begin{pmatrix} \frac{3}{4} & \frac{1}{4} & 0 \\ 0 & \frac{1}{2} & \frac{1}{2} \\ \frac{1}{2} & \frac{1}{2} & 0 \end{pmatrix}, \qquad \mathbf{c}^{(1)} = \begin{pmatrix} 16 \\ 7 \\ 0 \end{pmatrix}.$$

The vector $\mathbf{v}^{(1)}$ is determined by (9.10), so we must first obtain the matrix $I - \theta P^{(1)}$ and calculate its inverse

$$\left[I - \theta P^{(1)} \right] = \frac{1}{8} \begin{pmatrix} 5 & -1 & 0 \\ 0 & 6 & -2 \\ -2 & -2 & 8 \end{pmatrix}, \quad \left[I - \theta P^{(1)} \right]^{-1} = \frac{1}{27} \begin{pmatrix} 44 & 8 & 2 \\ 4 & 40 & 10 \\ 12 & 12 & 30 \end{pmatrix}.$$

Then it is easy to find

$$\mathbf{v}^{(1)} = \frac{1}{27} \begin{pmatrix} 760 \\ 344 \\ 276 \end{pmatrix} = \begin{pmatrix} 28.148 \\ 12.741 \\ 10.222 \end{pmatrix}.$$

The next step is policy iteration, so we must compare the components of $\mathbf{v}^{(1)}$ with those of the vector $\min\{\mathbf{c} + \theta P \mathbf{v}^{(1)}\}$. The first component has two possible values of $c_1 + \theta \sum_j p_{1j} v_j^{(1)}$:

$$v_1^{(1)} \text{ and } 16 + \frac{1}{2}\left(\frac{1}{2}v_1^{(1)} + \frac{1}{2}v_3^{(1)}\right) = 25.593.$$

The second of these is smaller, indicating that we must change the first row of $P^{(1)}$. Similarly, the two possible values of $c_2 + \theta \sum_j p_{2j} v_j^{(1)}$ are

$$v_2^{(1)} \text{ and } 4 + \frac{1}{2}\left(\frac{1}{2}v_1^{(1)} + \frac{1}{2}v_2^{(1)}\right) = 14.222.$$

Here $v_2^{(1)}$ gives the minimum, so we leave the second row of $P^{(1)}$ unchanged. Finally, the two alternative choices for $c_3 + \theta \sum_j p_{3j} v_j^{(1)}$ are

$$v_3^{(1)} \text{ and } 4 + \frac{1}{2}\left(\frac{1}{2}v_2^{(1)} + \frac{1}{2}v_3^{(1)}\right) = 9.741.$$

The minimum is obtained from the second choice, so we must change the third row of $P^{(1)}$ The result of all these detailed comparisons is the new policy matrix

$$P^{(2)} = \begin{pmatrix} \frac{1}{2} & 0 & \frac{1}{2} \\ 0 & \frac{1}{2} & \frac{1}{2} \\ 0 & \frac{1}{2} & \frac{1}{2} \end{pmatrix}, \qquad \mathbf{c}^{(2)} = \begin{pmatrix} 16 \\ 7 \\ 4 \end{pmatrix}.$$

Then calculations similar to those carried out for $P^{(1)}$ lead to the matrices

$$\left[I - \theta P^{(2)}\right] = \frac{1}{4}\begin{pmatrix} 3 & 0 & -1 \\ 0 & 3 & -1 \\ 0 & -1 & 3 \end{pmatrix}, \qquad \left[I - \theta P^{(2)}\right]^{-1} = \frac{1}{6}\begin{pmatrix} 8 & 1 & 3 \\ 0 & 9 & 3 \\ 0 & 3 & 9 \end{pmatrix}.$$

It turns out that

$$\mathbf{v}^{(2)} = \frac{1}{6} \begin{pmatrix} 147 \\ 75 \\ 57 \end{pmatrix} = \begin{pmatrix} 24.5 \\ 12.5 \\ 9.5 \end{pmatrix}.$$

A comparison of $\mathbf{v}^{(2)}$ with $\mathbf{v}^{(1)}$ shows that the policy $P^{(2)}$ is a substantial improvement on $P^{(1)}$. We now carry out a further policy iteration, starting with the vector $\mathbf{v}^{(2)}$. It is a straightforward matter to check that the only change occurs in the third row of $P^{(2)}$. Thus

$$P^{(3)} = \begin{pmatrix} \frac{1}{2} & 0 & \frac{1}{2} \\ 0 & \frac{1}{2} & \frac{1}{2} \\ \frac{1}{2} & \frac{1}{2} & 0 \end{pmatrix}, \qquad \mathbf{c}^{(3)} = \begin{pmatrix} 16 \\ 7 \\ 0 \end{pmatrix}.$$

We can evaluate $\mathbf{v}^{(3)}$ by standard calculations:

$$\left[I - \theta P^{(3)} \right] = \frac{1}{4} \begin{pmatrix} 3 & 0 & -1 \\ 0 & 3 & -1 \\ -1 & -1 & 4 \end{pmatrix},$$

$$\left[I - \theta P^{(3)} \right]^{-1} = \frac{1}{15} \begin{pmatrix} 22 & 2 & 6 \\ 2 & 22 & 6 \\ 6 & 6 & 18 \end{pmatrix},$$

$$\mathbf{v}^{(3)} = \frac{1}{15} \begin{pmatrix} 366 \\ 186 \\ 138 \end{pmatrix} = \begin{pmatrix} 24.4 \\ 12.4 \\ 9.2 \end{pmatrix}.$$

It is easy to confirm that the policy iteration procedure produces no further change of policy. In other words, $\mathbf{v}^{(3)}$ is the solution of the optimality equation, and hence the stationary policy determined by $P^{(3)}$ is optimal.

9.4   A MACHINE REPLACEMENT MODEL

So far we have been concerned with Markov decision processes having a finite number of states but, as we remarked earlier, many of the results can be extended to systems with infinitely many states. The problem we shall now investigate has the state space $S = \{0, 1, 2, \ldots \}$, and each $i \in S$ represents the performance level of a machine. The model is a stochastic version of the deterministic replacement problem mentioned in Chapter 4. The more general model here was first investigated in an early book on stochastic optimization by Ross [Ros70].

Imagine a production line where there is a particular type of machine in daily use. At the start of each day, the state i of the machine is observed and a decision is made whether to continue operating with the same machine or to replace it with a new one. We suppose there is an unlimited supply of new machines and that replacement of the current machine is instantaneous. The state of a new machine is always $i = 0$ and, roughly speaking, the performance of any machine deteriorates as i increases. This will be made more precise by

suitable assumptions about the running costs and transition probabilities at
different performance levels.

Let c_i be the cost of operating a machine for one day when it begins in
state i and let p_{ij} be the probability of a transition to state j by the start
of the next day. For any given state $i \geq 0$, there are two possible actions:
continue or replace. A decision to replace the current machine incurs a fixed
replacement cost $R > 0$ and the effect of this action is an instantaneous
transition to state zero. The choice is between a transition from state i
according to the probabilities p_{ij}, $j = 0, 1, 2, \ldots$ at cost c_i, or switching to a
new machine having transition probabilities p_{0j} at cost $R + c_0$. Future costs
will be discounted, using a factor θ with $0 < \theta < 1$. Let u_i be the expectation
of the total discounted cost over an infinite future, starting with a machine in
state i and using an optimal replacement policy. The minimum expected cost
function u satisfies the optimality equation

$$u_i = \min \left\{ R + c_0 + \theta \sum_j p_{0j} u_j, c_i + \theta \sum_j p_{ij} u_j \right\}, \qquad (9.11)$$

for each $i \geq 0$. This is similar to the optimality equation (9.4), except that
we are now dealing with an infinite vector $u = (u_0, u_1, u_2, \ldots)$ and the matrix
notation of the previous sections is no longer helpful.

The results of Propositions 9.1 and 9.2 are still valid. Let $x_i(n)$ be the
minimum expected discounted cost over a period of length n, given an initial
state $i \geq 0$. Then

$$x_i(n) = \min \left\{ R + c_0 + \theta \sum_j p_{0j} x_j(n-1), c_i + \theta \sum_j p_{ij} x_j(n-1) \right\}, \tag{9.12}$$

for $n \geq 1$. If all the running costs are non-negative, it is easy to show that, in
general,

$$0 \leq x_i(n) \leq x_i(n+1) \leq (R + c_0) / (1 - \theta).$$

The upper bound here is obtained by considering the policy of replacing the
existing machine by a new one every day. Then the argument of Proposition
9.1 shows that $x_i(n) \rightarrow u_i$ as $n \rightarrow \infty$, where the limit function $u =
(u_0, u_1, u_2, \ldots)$ is a solution of equation (9.11). Uniqueness is established by
a modification of the technique used in Proposition 9.2. It can be shown that
u is the only solution of the optimality equation which is non-negative and
bounded above.

Two further conditions will be needed. They reflect the idea that the
effectiveness of a machine is a decreasing function of its state. We shall rely

on the following assumptions

(i) $0 \leq c_0 \leq c_1 \leq c_2 \leq \cdots$.

(ii) For each integer $k \geq 0$, the sum $\sum_{j=k}^{\infty} p_{ij}$ is non-decreasing in i.

Proposition 9.6 *Let $h = (h_0, h_1, h_2, \ldots)$ be any non-negative function which is non-decreasing and bounded above. Assumption (ii) implies that*

$$\sum_{j=0}^{\infty} p_{ij} h_j \text{ is non-decreasing in } i.$$

Proof For each $i, k \geq 0$, let

$$q_{ik} = \sum_{j=k}^{\infty} p_{ij}, \text{ so that } p_{ik} = q_{ik} - q_{i,k+1}.$$

The series $\sum p_{ij} h_j$ is convergent, since the h_j are bounded. Hence, it can be rearranged in the following way.

$$\begin{aligned}
\sum_{j=0}^{\infty} p_{ij} h_j &= \sum_{j=0}^{\infty} (q_{ij} - q_{i,j+1}) h_j \\
&= q_{i0} h_0 + q_{i1} (h_1 - h_0) + q_{i2} (h_2 - h_1) + \cdots
\end{aligned}$$

All the terms of the new series are non-negative, since $0 \leq h_0 \leq h_1 \leq h_2 \leq \cdots$. Assumption (ii) means that each of the coefficients q_{ik} is non-decreasing in i, so the required result follows.   □

Proposition 9.6 will be used to show that the solution of the optimality equation (9.11) is a non-decreasing function. This fact leads to the conclusion that the optimal replacement policy must have a special form.

Proposition 9.7 *Let $u = (u_0, u_1, u_2, \ldots)$ be the unique solution of the optimality equation*

$$u_i = \min \left\{ R + c_0 + \theta \sum_j p_{0j} u_j, c_i + \theta \sum_j p_{ij} u_j \right\} \qquad (i \geq 0)$$

and assume that the conditions (i) and (ii) are satisfied. Then

$$0 \leq u_0 \leq u_1 \leq u_2 \leq \cdots .$$

Define the replacement level r by

$$r = \min \left\{ i \geq 0 : R + c_0 + \theta \sum_j p_{0j} u_j \leq c_i + \theta \sum_j p_{ij} u_j \right\}$$

if this is finite. Otherwise, set $r = \infty$. Then the policy of replacing the current machine as soon as it reaches a state $j \geq r$ is optimal.

Proof Consider the successive approximations determined by relation (9.12). We have $x_i(1) = \min\{R + c_0, c_i\}$ and assumption (i) guarantees that this is non-decreasing in i. Now suppose that the function

$$x(n-1) = (x_0(n-1), x_1(n-1), \dots)$$

is non-decreasing. We showed earlier that all the successive approximations are non-negative and bounded. Then assumptions (i) and (ii), together with Proposition 9.6, ensure that $c_i + \theta \sum_j p_{ij} x_j(n-1)$ is non-decreasing in i. Since the first expression on the right of (9.12) does not depend on i, it follows that $x(n)$ is a non-decreasing function. For each $n \geq 1$, we have

$$0 \leq x_0(n) \leq x_1(n) \leq x_2(n) \leq \cdots$$

In the limit when $n \to \infty$ and $x_i(n) \to u_i$, we obtain $0 \leq u_0 \leq u_1 \leq u_2 \leq \cdots$, as required.

Now consider the choice between the two possible actions: continue or replace, for a machine in state i. There is a definite advantage in continuing if

$$c_i + \theta \sum_j p_{ij} u_j < R + c_0 + \theta \sum_j p_{0j} u_j.$$

This is certainly true for $i = 0$, since R is positive. As i increases, so does the value of $c_i + \theta \sum_j p_{ij} u_j$. If the inequality remains true for every $i \geq 0$, we have $r = \infty$ and a replacement is never worthwhile. Otherwise, r is finite and

$$c_i + \theta \sum_j p_{ij} u_j \geq R + c_0 + \theta \sum_j p_0 u_j$$

holds for all $i \geq r$. The optimal action is replacement if $i \geq r$. In other words, the solution of the optimality equation determines a stationary policy with the replacement level r. The proof of Proposition 9.3 needs modification to allow for the infinite state space, but it is not difficult to confirm that this policy is optimal.   □

Proposition 9.7 shows that the solution of the optimality equation and the corresponding policy have a simple structure determined by the replacement level r. The value of this critical level can be found by solving the optimality equation, but this may involve substantial calculations.

Example 9.3 We are given the replacement cost $R > 0$ and a discount factor satisfying $0 < \theta < 1$. Suppose that the daily running cost increases linearly with the state of the machine: let $c_i = i$, for $i \geq 0$. The transition probabilities are defined as follows:

$$p_{i,i+1} = p, \qquad p_{ii} = q \qquad (i \geq 0),$$

where $q = 1 - p$ and $0 < p \leq 1$. Thus, we have a transition matrix determined by the single parameter p which is the probability of a change of state. It is easily verified that assumptions (i) and (ii) are satisfied. Note that the deterministic model used in Chapter 4 is included as a special case, with $p = 1$.

For some integer $r \geq 1$, the minimum expected cost function $u = (u_0, u_1, u_2, \dots)$ is determined by the system of equations

$$\begin{aligned}
u_i &= i + \theta\,(pu_{i+1} + qu_i) & (0 \leq i \leq r - 1),\\
u_i &= R + \theta\,(pu_1 + qu_0) & (i \geq r).
\end{aligned}$$

The critical level r will be determined later by using optimality conditions but, for the moment, let us treat r as fixed. It is convenient to introduce two new parameters

$$\lambda = \frac{1 - \theta q}{\theta p} \quad \text{and} \quad \mu = \frac{1}{\theta p}.$$

Note that $\mu > 1$ and, since $\lambda - 1 = \mu\,(1 - \theta)$, we also have $\lambda > 1$. The equation

$$u_0 = \theta\,(pu_1 + qu_0)$$

is equivalent to $u_1 = \lambda u_0$. The equations satisfied by the expected future costs can now be written in the form

$$u_{i+1} = \lambda u_i - \mu i, \tag{9.13}$$

for $0 \leq i \leq r - 1$, and $u_i = R + u_0$, for $i \geq r$.

Each u_i can be expressed in terms of u_0 by applying (9.13) for $i = 1, 2, \dots$. We find that

$$u_1 = \lambda u_0, \qquad u_2 = \lambda u_1 - \mu = \lambda^2 u_0 - \mu,$$

and so on. It is a straightforward matter to show that the solution of the equations (9.13) is

$$u_i = \lambda^i u_0 - \mu\left\{\lambda^i - (\lambda - 1)\,i - 1\right\}(\lambda - 1)^{-2}. \tag{9.14}$$

This formula is valid for $i = 1, 2, \dots, r$. Then the fact that $u_r = R + u_0$ leads to an expression for u_0. We find that

$$u_0 = \frac{\mu}{(\lambda - 1)^2} + \left\{R - \frac{\mu r}{(\lambda - 1)}\right\}\frac{1}{\lambda^r - 1}. \tag{9.15}$$

So far, the results apply to any policy which replaces the machine currently in use as soon as it reaches a state $j \geq r$. We must now find the optimal choice of r. According to Proposition 9.7, the solution of the optimality equation is a non-decreasing function and, by using (9.14), it is not difficult to check that $u_{i+1} \geq u_i$ if and only if

$$u_0 \geq \frac{\mu}{(\lambda - 1)^2} \left\{ 1 - \frac{1}{\lambda^i} \right\}.$$

The expression on the right here is increasing in i, so the inequality holds for all $i \leq r - 1$ if it holds for $i = r - 1$. In other words, the condition

$$u_0 \geq \frac{\mu}{(\lambda - 1)^2} \left\{ 1 - \frac{1}{\lambda^{r-1}} \right\} \tag{9.16}$$

is enough to ensure that $u_0 \leq u_1 \leq \cdots u_r$. By using (9.15) to eliminate u_0 and by rearranging the terms in the resulting inequality, it can be shown that (9.16) is equivalent to

$$\frac{1}{\lambda^{r-1}} + (\lambda - 1)(r - 1) \leq \frac{(\lambda - 1)^2}{\mu} R + 1. \tag{9.17}$$

The replacement level r must be chosen to satisfy this optimality condition and we need to impose one further condition to guarantee that the replacement policy is optimal.

For this example, the optimality equations can be written in the form

$$u_i = \min \left\{ R + u_0, i + \theta \left(p u_{i+1} + q u_i \right) \right\}. \tag{9.18}$$

We have constructed a solution, given by (9.14) and (9.15), which satisfies these equations for $i = 0, 1, \ldots, r - 1$, provided that r is chosen according to condition (9.17). The solution is given by $u_i = R + u_0$ for $i \geq r$, so the remaining optimality equations are valid if

$$R + u_0 \leq i + \theta \left(p u_{i+1} + q u_i \right).$$

This reduces to

$$R + u_0 \leq i + \theta \left(R + u_0 \right),$$

because the value of u_i remains constant for $i \geq r$. Hence, we can ensure that $u = (u_0, u_1, u_2, \ldots)$ is the unique solution of the optimality equations by demanding that

$$(R + u_0)(1 - \theta) \leq r.$$

By using the fact that $(1 - \theta) = (\lambda - 1)/\mu$ and equation (9.15) for u_0, this condition can be expressed in a form comparable with (9.17). After some awkward rearrangements, we obtain the final optimality condition

$$\frac{1}{\lambda^r} + (\lambda - 1) r \geq \frac{(\lambda - 1)^2}{\mu} R + 1. \tag{9.19}$$

We conclude this analysis of the model specified in Example 9.3 by noting that the inequalities (9.17) and (9.19) always determine an optimal replacement level $r \geq 1$. It is easy to see that, since $\lambda > 1$, the expression $\lambda^{-j} + (\lambda - 1) j$ is strictly increasing in j for $j = 1, 2, \ldots$ and it has no upper bound. Hence, for any replacement cost $R > 0$ and any parameters $\lambda, \mu > 1$, we can define r as the smallest positive integer satisfying condition (9.19). Then condition (9.17) is also satisfied and the replacement policy with the critical level r is optimal. If equality happens to occur in (9.19) for the integer r, then $r + 1$ is also an optimal replacement level. Finally, it is worth noting that r can be chosen by minimizing the total discounted expectation u_0 given by (9.15). This leads to the same optimality conditions (9.17) and (9.19).

10

Minimizing Average Costs

10.1 INTRODUCTION

This chapter is concerned with the problem of minimizing the long-term average costs for a Markov decision process. The model and notation are similar to those of the previous chapter, except that future costs will not be discounted: $\theta = 1$ here. As before, we have a finite state space $S = \{1, 2, \ldots, s\}$ and the motion of the system is controlled by choosing a transition matrix $P \in D$ at each point of discrete time. The set D of admissible transition matrices is finite and the choice at each stage is made by treating the rows of P separately. In other words, for any current state $j \in S$, the selection of a probability distribution $\mathbf{p}_j = (p_{j1}, p_{j2}, \ldots, p_{js})$ for the next transition determines the immediate cost c_j of the transition and this does not affect the possible choices for any other row of P or the corresponding cost. There is a column vector $\mathbf{c} = (c_1, c_2, \ldots, c_s)$ associated with each transition matrix $P \in D$ which gives the immediate costs determined by the rows of P.

Let us assume that all the transition costs satisfy

$$0 \le c_j \le b, \tag{10.1}$$

for some constant b. Consider the sequence of vectors $\{\mathbf{x}(n)\}$, where $x_i(n)$ is the minimum expected cost over n transitions, starting in state i. These vectors are determined by relation (9.3) with $\theta = 1$. We have $\mathbf{x}(0) = \mathbf{0}$ and

$$\mathbf{x}(n) = \min\{\mathbf{c} + P\mathbf{x}(n-1)\}, \tag{10.2}$$

for $n \ge 1$. This minimization operation has the same interpretation as before, with each component treated separately:

$$x_i(n) = \min\left\{c_i + \sum_j p_{ij} x_j(n-1)\right\} \qquad (1 \le i \le s)$$

and the expression on the right-hand side is minimized with respect to $\mathbf{p}_i = (p_{i1}, p_{i2}, \ldots, p_{is})$ and the corresponding cost c_i. The sequence $\{\mathbf{x}(n)\}$

is non-decreasing, in view of assumption (10.1), but it usually diverges since there is no discounting of future costs. The asymptotic behaviour of $\mathbf{x}(n)$ as $n \to \infty$ can be illustrated by looking at the special case when there is only one admissible matrix P. Then the underlying random process is a Markov chain and relation (10.2) reduces to $\mathbf{x}(n) = \mathbf{c} + P\mathbf{x}(n-1)$. Hence

$$\mathbf{x}(n) = \mathbf{c} + P\mathbf{c} + \cdots + P^{n-1}\mathbf{c}$$

for each $n \geq 1$. The sequence $\{\mathbf{x}(n)\}$ may exhibit periodic behaviour if the underlying Markov chain is periodic.

Example 10.1 Let $S = \{1,2\}$ and suppose that the only admissible transition matrix P and its cost vector $\mathbf{c}$ are given by

$$P = \begin{pmatrix} 0 & 1 \\ 1 & 0 \end{pmatrix}, \qquad\qquad \mathbf{c} = \begin{pmatrix} 1 \\ 0 \end{pmatrix}.$$

In this case, P^2 is the identity matrix and hence,

$$P^n = I \text{ if } n \text{ is even}, \qquad P^n = P \text{ if } n \text{ is odd}.$$

The sequence $\{\mathbf{x}(n)\}$ begins with

$$\mathbf{x}(1) = \mathbf{c} = \begin{pmatrix} 1 \\ 0 \end{pmatrix}, \quad \mathbf{x}(2) = \mathbf{c} + P\mathbf{c} = \begin{pmatrix} 1 \\ 0 \end{pmatrix} + \begin{pmatrix} 0 \\ 1 \end{pmatrix} = \begin{pmatrix} 1 \\ 1 \end{pmatrix}.$$

We have

$$\mathbf{x}(n) = \mathbf{c} + P\mathbf{c} + \mathbf{c} + P\mathbf{c} + \cdots + P^{n-1}\mathbf{c},$$

and the final term is $\mathbf{c}$ or $P\mathbf{c}$, according to whether n is odd or even. It is easily verified that

$$\mathbf{x}(n) = \frac{1}{2}n \begin{pmatrix} 1 \\ 1 \end{pmatrix} + \frac{1}{4}(1-(-1)^n) \begin{pmatrix} 1 \\ -1 \end{pmatrix} \qquad\qquad (n \geq 1).$$

Each component $\mathbf{x}_i(n) \to \infty$ as $n \to \infty$, but $\mathbf{x}(n)$ also contains a periodic term with the same period 2 as the transition matrix P. In what follows, we shall be mainly concerned with the limiting behaviour of the average expected cost $\mathbf{x}(n)/n$. For this example, it is clear that

$$\frac{1}{n}\mathbf{x}(n) \to \frac{1}{2} \begin{pmatrix} 1 \\ 1 \end{pmatrix} \text{ as } n \to \infty.$$

In general, the minimum expected cost vectors $\mathbf{x}(n)$ for a finite Markov decision process are asymptotically periodic, but it has been proved that the average cost $\mathbf{x}(n)/n$ always converges to a limit vector as $n \to \infty$. The relation

(10.2) generates a policy sequence. Let $P^{(n)} \in D$ be a transition matrix which, together with its immediate cost vector $\mathbf{c}^{(n)}$, attains the minimum in (10.2):

$$\mathbf{x}(n) = \mathbf{c}^{(n)} + P^{(n)}\mathbf{x}(n-1) \qquad\qquad (n \geq 1).$$

The policy sequence $\{P^{(n)}\}$ need not converge and its behaviour may be quite irregular. An example has been constructed in which $P^{(n)}$ oscillates as n increases, but never becomes periodic: see Part I of the series of papers by Bather [Bat73]. This paper also gives a brief summary of the early development of Markov decision theory. A detailed description of the historical background and the broad range of results now available on Markov decision processes can be found in the recent book by Puterman [Put94].

The aim here is to concentrate on basic results which can be obtained and understood without introducing more specialized mathematical tools. One way of reducing the possible complications is to focus on finite systems which are *communicating* in the sense that each state $j \in S$ is accessible from any initial state $i \in S$, by using a suitable policy sequence. In this case, the minimum average cost does not depend on the initial state:

$$\frac{1}{n}\mathbf{x}(n) \to \gamma \mathbf{1} \qquad\qquad \text{as } n \to \infty.$$

In other words, $\mathbf{x}_i(n)/n \to \gamma$ for any initial state i and the constant γ represents the minimum long-term average cost for the Markov decision process. Communicating systems will be investigated in Section 10.2. The optimality equation takes the form:

$$\boldsymbol{\xi} + \gamma \mathbf{1} = \min\{\mathbf{c} + P\boldsymbol{\xi}\}. \qquad\qquad (10.3)$$

It will be shown that, if we can find a vector $\boldsymbol{\xi}$ and a constant γ which satisfy this equation, then γ is the minimum average cost for the system and any transition matrix $P \in D$ which attains the minimum on the right determines an optimal policy.

This result is extended to systems with infinitely many states in Section 10.3. It turns out that a solution of the optimality equation:

$$\xi_i + \gamma = \min\left\{c_i + \sum_j p_{ij}\xi_j\right\}$$

determines an optimal policy, provided that the set of values $\{\xi_i, i \in S\}$ remains bounded. The machine replacement model of Chapter 9 will be used to illustrate this. The optimal policy in Example 9.3 has a limiting form as the discount factor $\theta \to 1$ and this limiting policy is optimal with respect to the average cost criterion when $\theta = 1$.

Section 10.4. is concerned with a classical problem of inventory theory. This has stimulated a good deal of research, including a recent paper which shows how an optimal policy can be justified even when it corresponds to an unbounded solution of the optimality equation.

10.2   LONG-TERM AVERAGE COSTS

A brief introduction to Markov chains was given in Chapter 7. Here, it will be useful to consider their long term behaviour. Let P be the transition matrix of a finite Markov chain with the state space $S = \{1, 2, \ldots, s\}$. We recall that the n-step transition matrix is P^n. In general, P^n does not converge to a limit as $n \to \infty$. Periodic behaviour may occur, as in Example 10.1. However, a limit can always be obtained by averaging over time. This is called the Cesaro limit. Let

$$A^{(n)} = \frac{1}{n} \left[I + P + \cdots + P^{n-1} \right]$$

and note that $A^{(n)}$ is a stochastic matrix for each $n \geq 1$. Then $A^{(n)}$ converges to a limit matrix P^* as $n \to \infty$. In other words, each element $a_{ij}^{(n)} \to p_{ij}{}^*$ as $n \to \infty$. It follows that $P^* = (p_{ij}{}^*)$ is also a stochastic matrix. Proofs of this and other properties stated below can be found in Kemeny and Snell [KS60]. Similar properties hold for countable Markov chains and there is a more comprehensive treatment of results relevant to Markov decision theory in the appendix of Puterman's book [Put94].

Any finite stochastic matrix P has the following properties:

(i) $\dfrac{1}{n} \left[I + P + \cdots + P^{n-1} \right] \to P^*$ as $n \to \infty$.

(ii)   The limit P^* is a stochastic matrix such that

$$PP^* = P^*P = P^*.$$

(iii) The matrix $[I - P + P^*]$ is non-singular.

(iv)   If P^n converges as $n \to \infty$, then the limit is P^*.

We now return to the Markov decision process described in the previous section and consider a stationary policy determined by a pair $(P, \mathbf{c})$, where $P \in D$ and $\mathbf{c}$ is the associated cost vector. The total expected cost over n transitions is represented by the vector

$$\mathbf{c} + P\mathbf{c} + \cdots + P^{n-1}\mathbf{c}$$

and the corresponding average cost is

$$\frac{1}{n} \left[I + P + \cdots + P^{n-1} \right] \mathbf{c}.$$

This converges to $P^*\mathbf{c}$ as $n \to \infty$, so our problem is to choose $P \in D$ in order to minimize the components of the limit vector $\boldsymbol{\eta} = P^*\mathbf{c}$. Each component η_i

is the long-term average expected cost for the stationary policy P when the initial state is i.

For some transition matrices, the long-term average cost does not depend on the initial state. For example, this occurs if the Markov chain determined by P is irreducible, but it is difficult to guarantee this for every choice of $P \in D$ without making very restrictive assumptions. Our aim is to find a policy which attains the minimum average cost for every state and there are relatively simple conditions which guarantee that this minimum is the same for any initial state.

Definition 10.1 *We say that a finite Markov decision process is a communicating system if, for any pair of distinct states $i, j \in S$, there is a policy $P \in D$ and a positive integer r such that the r-step transition probability $p_{ij}^{(r)} > 0$.*

From now on, it will be assumed that we have a communicating system. Thus, each state j is accessible from any other state i by using a suitable policy. In the special case when there is only one admissible transition matrix $P \in D$, the Markov decision process reduces to a Markov chain and the assumption means that P is irreducible.

The main consequence of assuming that we have a communicating system is the existence of a vector $\boldsymbol{\xi}$ and a constant γ which satisfy the optimality equation (10.3). This result is not easy to establish: for a proof, see Part II of the papers by Bather [Bat73]. However, any solution of the optimality equation leads in a straightforward way to an optimal policy.

Proposition 10.1 *Suppose that $\boldsymbol{\xi}$ is a vector such that*

$$\boldsymbol{\xi} + \gamma \mathbf{1} = \min_{P \in D} \{\mathbf{c} + P\boldsymbol{\xi}\}$$

for some constant γ. Suppose further that the minimum on the right is attained by a pair $(Q, \mathbf{d})$, where $Q \in D$ and $\mathbf{d}$ is the corresponding immediate cost vector. Then

$$Q^*\mathbf{d} = \gamma \mathbf{1} \text{ and } P^*\mathbf{c} \geq \gamma \mathbf{1}$$

for any other admissible pair $(P, \mathbf{c})$. Hence, the stationary policy determined by Q is optimal and it attains the minimum long-term average cost γ, for any initial state.

Proof We have

$$\boldsymbol{\xi} + \gamma \mathbf{1} = \mathbf{d} + Q\boldsymbol{\xi}$$

but, for any other pair $(P, \mathbf{c})$,

$$\boldsymbol{\xi} + \gamma \mathbf{1} \leq \mathbf{c} + P\boldsymbol{\xi}.$$

Starting with this inequality, we obtain

$$\boldsymbol{\xi} \leq \mathbf{c} + P\boldsymbol{\xi} - \gamma\mathbf{1}, \qquad P\boldsymbol{\xi} \leq P\mathbf{c} + P^2\boldsymbol{\xi} - \gamma\mathbf{1},$$

since $P\mathbf{1} = \mathbf{1}$. Hence,

$$P^n\boldsymbol{\xi} \leq P^n\mathbf{c} + P^{n+1}\boldsymbol{\xi} - \gamma\mathbf{1}$$

for each $n \geq 1$. By using this, it is easy to verify inductively that

$$\boldsymbol{\xi} \leq \mathbf{c} + P\mathbf{c} + \cdots + P^{n-1}\mathbf{c} + P^n\boldsymbol{\xi} - n\gamma\mathbf{1}.$$

We have shown that, for each $n \geq 1$,

$$\frac{1}{n}\left[I + P + \cdots + P^{n-1}\right]\mathbf{c} \geq \frac{1}{n}\left[I - P^n\right]\boldsymbol{\xi} + \gamma\mathbf{1}.$$

As $n \to \infty$, the left-hand side converges to $P^*\mathbf{c}$ and the first term on the right converges to $\mathbf{0}$, so we may conclude that $P^*\mathbf{c} \geq \gamma\mathbf{1}$.

The above argument can be repeated, starting with the equation

$$\boldsymbol{\xi} + \gamma\mathbf{1} = \mathbf{d} + Q\boldsymbol{\xi}.$$

It leads to the relation

$$\boldsymbol{\xi} = \mathbf{d} + Q\mathbf{d} + \cdots + Q^{n-1}\mathbf{d} + Q^n\boldsymbol{\xi} - n\gamma\mathbf{1},$$

for $n \geq 1$, and finally to the result that $Q^*\mathbf{d} = \gamma\mathbf{1}$, as required. □

Let $\{\mathbf{x}(n)\}$ be the sequence of minimum expected cost vectors generated by relation (10.2). It is not difficult to show, under the above conditions, that the minimum average cost $x_i(n)/n$ over a finite period converges to γ as $n \to \infty$, for any initial state $i \in S$. Notice first that, if $\boldsymbol{\xi}$ and γ satisfy the optimality equation, then the vector $\boldsymbol{\xi}$ can be replaced by $\boldsymbol{\xi} + \lambda\mathbf{1}$ for any constant λ, since $P(\lambda\mathbf{1}) = \lambda\mathbf{1}$ for any choice of transition matrix $P \in D$. The solution vector $\boldsymbol{\xi}$ is not unique, but it is clear from Proposition 10.1 that γ is unique.

Proposition 10.2 *Suppose that*

$$\boldsymbol{\xi} + \gamma\mathbf{1} = \min\{\mathbf{c} + P\boldsymbol{\xi}\},$$

as before, and consider the minimum expected cost vectors defined by $\mathbf{x}(0) = \mathbf{0}$ *and*

$$\mathbf{x}(n) = \min\{\mathbf{c} + P\mathbf{x}(n-1)\}$$

for $n \geq 1$. Then

$$\frac{1}{n}\mathbf{x}(n) \to \gamma\mathbf{1} \text{ as } n \to \infty.$$

Proof Choose constants α and β such that

$$\alpha \leq \xi_i \leq \beta \qquad (i \in S).$$

Then the vectors $\boldsymbol{\xi} - \alpha\mathbf{1}$ and $\boldsymbol{\xi} - \beta\mathbf{1}$ both satisfy the optimality equation and we have

$$\boldsymbol{\xi} - \alpha\mathbf{1} \geq \mathbf{x}(0) \geq \boldsymbol{\xi} - \beta\mathbf{1}.$$

It will be established that, for $n \geq 1$,

$$\boldsymbol{\xi} - \alpha\mathbf{1} + n\gamma\mathbf{1} \geq \mathbf{x}(n) \geq \boldsymbol{\xi} - \beta\mathbf{1} + n\gamma\mathbf{1}. \qquad (10.4)$$

For example, the vector $\boldsymbol{\xi} - \beta\mathbf{1}$ satisfies

$$(\boldsymbol{\xi} - \beta\mathbf{1}) + \gamma\mathbf{1} = \min\{\mathbf{c} + P(\boldsymbol{\xi} - \beta\mathbf{1})\}.$$

Since $\boldsymbol{\xi} - \beta\mathbf{1} \leq 0$, we have

$$\mathbf{x}(1) = \min\{\mathbf{c} + P\mathbf{0}\} \geq \min\{\mathbf{c} + P(\boldsymbol{\xi} - \beta\mathbf{1})\},$$

so $\mathbf{x}(1) \geq \boldsymbol{\xi} - \beta\mathbf{1} + \gamma\mathbf{1}$. The inductive hypothesis that

$$\mathbf{x}(n-1) \geq \boldsymbol{\xi} - \beta\mathbf{1} + (n-1)\gamma\mathbf{1}$$

leads to the conclusion that

$$\begin{aligned}
\mathbf{x}(n) &= \min\{\mathbf{c} + P\mathbf{x}(n-1)\} \\
&\geq \min\{\mathbf{c} + P(\boldsymbol{\xi} - \beta\mathbf{1})\} + (n-1)\gamma\mathbf{1} \\
&= (\boldsymbol{\xi} - \beta\mathbf{1}) + n\gamma\mathbf{1},
\end{aligned}$$

by using the optimality equation for $\boldsymbol{\xi} - \beta\mathbf{1}$.

A similar inductive argument, starting with the vector $\boldsymbol{\xi} - \alpha\mathbf{1}$, can be used to verify that

$$\mathbf{x}(n) \leq (\boldsymbol{\xi} - \alpha\mathbf{1}) + n\gamma\mathbf{1}$$

always holds. Then the inequalities (10.4) show that

$$\frac{1}{n}(\boldsymbol{\xi} - \alpha\mathbf{1}) + \gamma\mathbf{1} \geq \frac{1}{n}\mathbf{x}(n) \geq \frac{1}{n}(\boldsymbol{\xi} - \beta\mathbf{1}) + \gamma\mathbf{1},$$

and it follows that

$$\frac{1}{n}\mathbf{x}(n) \to \gamma\mathbf{1} \quad \text{as} \quad n \to \infty,$$

as required. $\qquad\qquad\qquad\qquad\qquad\qquad\qquad\qquad\qquad\qquad\qquad\square$

Policy improvements
It is not easy to find solutions of the optimality equation. The method of
successive approximations used in the previous chapter is no longer effective.
Without discounting, the sequence of minimum expected cost vectors may
diverge and it may also be asymptotically periodic. The most useful method
of constructing optimal policies is the policy iteration technique developed by
Howard [How60]. For average-optimal policies, it is preferable to rely on a
modified form of policy improvement due to Blackwell [Bla62]. This method
and its properties will be described briefly, without attempting any proofs.

For any finite stochastic matrix P, let

$$P^* = \lim_{n \to \infty} \frac{1}{n} \left[I + P + \cdots + P^{n-1} \right]. \tag{10.5}$$

As we noted in Section 10.2, the limit P^* always exists and

$$PP^* = P^*P = P^*.$$

The matrix $[I - P + P^*]$ is non-singular. Now let $P \in D$ and let $\mathbf{c}$ be its
immediate cost vector. We define two other vectors associated with P:

$$\boldsymbol{\eta} = P^*\mathbf{c}, \qquad \boldsymbol{\zeta} = [I - P + P^*]^{-1}[I - P^*]\mathbf{c}. \tag{10.6}$$

It can be shown that they have the following properties:

$$P\boldsymbol{\eta} = \boldsymbol{\eta}, \qquad P^*\boldsymbol{\zeta} = \mathbf{0}, \qquad \boldsymbol{\zeta} + \boldsymbol{\eta} = \mathbf{c} + P\boldsymbol{\zeta}. \tag{10.7}$$

The vector $\boldsymbol{\eta}$ represents the average cost attained by the policy P and both
vectors $\boldsymbol{\eta}$ and $\boldsymbol{\zeta}$ are used in policy improvements.

Consider a finite Markov decision process with state space $S = \{1, 2, \ldots, s\}$
and let D be the set of admissible policies. A policy $P \in D$ is selected by
choosing each row $\mathbf{p}_i = (p_{i1}, p_{i2}, \ldots, p_{is})$ from a finite set D_i, separately for
$i = 1, 2, \ldots, s$. We start with a stationary policy $P^{(1)} \in D$ and then determine
new policies $P^{(2)}, P^{(3)}, \ldots$ by successive improvements. For each $P^{(r)} \in D$,
let $\mathbf{c}^{(r)}$ be the prescribed cost vector, let $\boldsymbol{\eta}^{(r)}$ and $\boldsymbol{\zeta}^{(r)}$ be the corresponding
vectors defined by (10.6).

For any stationary policy $P^{(1)} \in D$, a new policy $P^{(2)} \in D$ is constructed
according to the following rules:

The ith row $P^{(2)}$ is selected from D_i so that either

(i) $\sum_j p_{ij}^{(2)} \eta_j^{(1)} < \sum_j p_{ij}^{(1)} \eta_j^{(1)}$

or the corresponding equality holds, together with

(ii) $c_i^{(2)} + \sum_j p_{ij}^{(2)} \zeta_j^{(1)} < c_i^{(1)} + \sum_j p_{ij}^{(1)} \zeta_j^{(1)}$.

If neither of these is possible, the ith row of $P^{(1)}$ must be retained in $P^{(2)}$, $i = 1, 2, \ldots, s$.

Except in the case when $P^{(2)} = P^{(1)}$, the new policy is an improvement in the sense described below.

In his paper, Blackwell [Bla62] established the consequences of these rules. For every state $i \in S$, the average cost

$$\eta_i^{(2)} \leq \eta_i^{(1)}$$

$$\text{and} \qquad \zeta_i^{(2)} \leq \zeta_i^{(1)} \qquad \text{whenever } \eta_i^{(2)} = \eta_i^{(1)}.$$

Because of the condition that no change in policy is made except where this is indicated in (i) or (ii) by a strict inequality, $\boldsymbol{\eta}^{(2)} = \boldsymbol{\eta}^{(1)}$ and $\boldsymbol{\zeta}^{(2)} = \boldsymbol{\zeta}^{(1)}$ cannot both hold unless $P^{(2)} = P^{(1)}$. Repeated iterations lead to a sequence of policies $P^{(1)}, P^{(2)}, \ldots, P^{(k)}$ and a decreasing sequence of average cost vectors:

$$\boldsymbol{\eta}^{(1)} \geq \boldsymbol{\eta}^{(2)} \geq \cdots \geq \boldsymbol{\eta}^{(k)}.$$

The iterations continue until no further improvement is possible, so that $P^{(k+1)} = P^{(k)}$. This must occur for some $k \geq 1$, since the set of admissible policies is finite. The final policy is optimal:

$$\boldsymbol{\eta}^{(k)} = \min_{P \in D} \{P^* \mathbf{c}\}.$$

These results are similar to those established in Section 9.3 for Markov decision processes with discounted future costs, but they are more difficult to prove. In the case when the system is communicating, the minimum average cost is the same for every initial state $i \in S$ and, for the optimal policy $P^{(k)}$,

$$\boldsymbol{\eta}^{(k)} = \gamma \mathbf{1}.$$

However, this does not apply to other policies in the sequence: the average cost vectors $\boldsymbol{\eta}^{(r)}$ may not have this special form for $r < k$.

10.3   EXTENSION TO INFINITELY MANY STATES

Consider a Markov decision process with the state space $S = \{0, 1, 2, \ldots\}$. Since there are infinitely many states, the vector notation used earlier is not so helpful. We will rely on components representing the conditional expectations of future costs, for given initial states. For any state $i \in S$, the probability distribution $p_i = (p_{i0}, p_{i1}, p_{i2}, \ldots)$ for the next transition is chosen from a finite set D_i and the corresponding cost is c_i. It will be assumed that

$$c_i \geq 0$$

always. As before, $x_i(n)$ is the minimum expected cost over n transitions, starting in state i. The vector relation (10.2) is replaced by

$$x_i(n) = \min \left\{ c_i + \sum_j p_{ij} x_j(n-1) \right\} \tag{10.8}$$

for each $i \in S$ and $n \geq 1$, with $x_j(0) = 0$ always. Since the costs c_i are non-negative, it is clear that $x_i(n) \geq 0$ and the sequence $\{x_i(n), n \geq 1\}$ is non-decreasing. Notice that the minimization in (10.8) may involve considering, and rejecting, cases $p_i \in D_i$ where the series $\sum_{j=0}^{\infty} p_{ij} x_j(n-1)$ diverges. This is not important, provided that the series converges for at least one $p_i \in D_i$. It will be assumed that this always holds, so that every minimum expected cost $x_i(n)$ is finite.

The optimality equation (10.3) now becomes an infinite system of equations

$$\xi_i + \gamma = \min \left\{ c_i + \sum_j p_{ij} \xi_j \right\} \tag{10.9}$$

for each $i \in S$. The solution, if it exists, is made up of a function $\xi = (\xi_0, \xi_1, \xi_2, \ldots)$ and a constant γ. Many of the results described earlier, for finite Markov decision processes, do not extend to systems with infinitely many states. For example, there may be no solution of the optimality equations even if the system is communicating. In some cases, there is a solution, but the function ξ is unbounded. Here, we will restrict attention to the situation where there is a bounded solution. Then the solution determines an optimal policy and results similar to Propositions 10.1 and 10.2 can be established.

Suppose that $\xi = (\xi_0, \xi_1, \xi_2, \ldots)$ and γ satisfy the optimality equation (10.9), and suppose further that

$$\alpha \leq \xi_i \leq \beta, \tag{10.10}$$

for $i \in S$, where α and β are constants. For each i, let $q_i = (q_{i0}, q_{i1}, q_{i2}, \ldots) \in D_i$ and the corresponding cost d_i attain the minimum in (10.9) and consider the stationary policy $Q = (q_{ij})$ determined in this way. We have

$$\xi_i + \gamma = d_i + \sum_j q_{ij} \xi_j \tag{10.11}$$

but, for other possible choices $p_i \in D_i$,

$$\xi_i + \gamma \leq c_i + \sum_j p_{ij} \xi_j. \tag{10.12}$$

For each initial state i, the expected cost using the stationary policy Q over a finite period is given by

$$y_i(n) = d_i + \sum_j q_{ij} y_j(n-1) \tag{10.13}$$

for $n \geq 1$, where $y_j(0) = 0$ always.

Proposition 10.3 *Suppose that the optimality equations have a bounded solution, satisfying the conditions (10.10). Let Q be a stationary policy constructed so that the equations (10.11) hold and let the corresponding expected future costs $y_i(n)$ be defined by the relations (10.13). Then, for each $i \in S$ and $n \geq 1$,*

$$\xi_i - \alpha + n\gamma \geq y_i(n) \geq x_i(n) \geq \xi_i - \beta + n\gamma.$$

Hence,

$$\lim_{n \to \infty} \frac{y_i(n)}{n} = \lim_{n \to \infty} \frac{x_i(n)}{n} = \gamma$$

and the policy Q is optimal.

Proof Since $x_i(n)$ is the minimum expected cost over n transitions, it is clear that $x_i(n) \leq y_i(n)$. An argument similar to that used in Proposition 10.2. shows that

$$x_i(n) \geq \xi_i - \beta + n\gamma.$$

It remains to prove that, in general,

$$y_i(n) \leq \xi_i - \alpha + n\gamma.$$

We have $y_i(0) = 0 \leq \xi_i - \alpha$ by (10.10), so let us assume that

$$y_j(n-1) \leq \xi_j - \alpha + (n-1)\gamma$$

for some $n \geq 2$ and all $j \in S$. Then, according to relation (10.13),

$$y_i(n) = d_i + \sum_j q_{ij} y_j(n-1) \leq d_i + \sum_j q_{ij}\xi_j - \alpha + (n-1)\gamma,$$

since $\sum_j q_{ij} = 1$. Finally, equation (10.11) can be used to reduce this to

$$y_i(n) \leq \xi_i - \alpha + n\gamma.$$

We have now established all the required inequalities, and they show that

$$\gamma + \frac{1}{n}(\xi_i - \alpha) \geq \frac{1}{n}y_i(n) \geq \frac{1}{n}x_i(n) \geq \gamma + \frac{1}{n}(\xi_i - \beta).$$

Hence the ratios $y_i(n)/n$ and $x_i(n)/n$ both converge to γ as $n \to \infty$.   □

The example which follows is based on the machine replacement model of Chapter 9. It is concerned with the limiting form of the optimal policy in Example 9.3. This will be justified, by applying Proposition 10.3, to show that it minimizes the long-term average cost.

Example 10.2 Consider the model of Section 9.4. Each state $i \geq 0$ represents the performance level of a machine. Let the operating cost for one day, starting in state i, be $c_i = i$, and suppose that the transition probabilities giving the distribution of the next state are

$$p_{i,i+1} = p, \qquad p_{ii} = q \qquad (i \geq 0),$$

where $0 < p \leq 1$, $q = 1 - p$ and $p_{ij} = 0$, otherwise. As before, the cost of replacing the current machine by a new one is $R > 0$ and this is equivalent to an instantaneous transition from state i to zero. There is no discounting here: $\theta = 1$, so the optimality equations are

$$\xi_i + \gamma = \min\{R + p\xi_1 + q\xi_0, i + p\xi_{i+1} + q\xi_i\}, \qquad (10.14)$$

for $i \geq 0$.

We proved that the optimal policy for the discounted version is based on a critical replacement level r and the rule:

replace the current machine as soon as it reaches a state $j \geq r$.

In Example 9.3, the optimal replacement level r is chosen to satisfy the inequalities (9.17) and (9.19). It is a straightforward matter to find the limiting form of these optimality conditions as $\theta \to 1$.

Let $\theta = 1 - \delta$ and consider the inequalities (9.17) and (9.19) as δ decreases to zero. Terms of order δ^2 are important, but smaller terms may be neglected. We first obtain

$$\lambda = \frac{1 - \theta q}{\theta p} = 1 + \frac{1}{p}\left(\delta + \delta^2\right) + o\left(\delta^2\right),$$

$$\mu = \frac{1}{\theta p} = \frac{1}{p}\left(1 + \delta + \delta^2\right) + o\left(\delta^2\right).$$

Similar calculations show that

$$\frac{(\lambda - 1)^2}{\mu} = \frac{1}{p}\delta^2 + o\left(\delta^2\right),$$

$$\frac{1}{\lambda^r} + (\lambda - 1)r = 1 + \frac{r(r+1)}{2p^2}\delta^2 + o\left(\delta^2\right).$$

It follows that the optimality condition (9.19) reduces to

$$1 + \frac{r(r+1)}{2p^2}\delta^2 \geq 1 + \frac{R}{p}\delta^2 + o\left(\delta^2\right).$$

By comparing the coefficients of δ^2 here, we can deduce that r satisfies this inequality as $\delta \to 0$ only if

$$r(r+1) \geq 2pR.$$

The condition (9.17) can be reduced in the same way to $r(r-1) \leq 2pR$. In other words, the limiting form of the optimal policy is determined by a critical replacement level r such that

$$r(r-1) \leq 2pR \leq r(r+1). \qquad (10.15)$$

From now on, let us assume that r is chosen to be the smallest positive integer satisfying both of these inequalities. It remains to verify that the replacement policy defined by this critical level corresponds to a bounded solution of the optimality equations (10.14).

The first of the optimality equations shows that

$$\xi_0 + \gamma = p\xi_1 + \xi_0,$$

since $R > 0$. For simplicity, let $\xi_0 = 0$ so that $\gamma = p\xi_1$. We seek a solution which corresponds to a policy with the replacement level r. The equations associated with this policy will enable us to construct a solution, leaving the optimality conditions to be verified later. For states $i \geq r$, the machine should be replaced, so we have $\xi_i + \gamma = R + p\xi_1$. Since $\gamma = p\xi_1$, this reduces to

$$\xi_i = R \qquad\qquad (i \geq r).$$

The current machine should continue to operate if $i \leq r - 1$ and the corresponding equation is

$$\xi_i + \gamma = i + p\xi_{i+1} + q\xi_i,$$

which leads to

$$p(\xi_{i+1} - \xi_i) = \gamma - i \qquad\qquad (0 \leq i \leq r - 1).$$

This set of equations can be solved without much difficulty, and the result is given by

$$\xi_i = \frac{1}{p}\left\{\gamma i - \frac{1}{2}i(i-1)\right\}, \qquad (10.16)$$

for $0 \leq i \leq r$. Then, since $\xi_r = R$, we can obtain a relation which determines the average cost γ. Hence,

$$\gamma = \frac{1}{2}(r-1) + \frac{pR}{r} \tag{10.17}$$

and by using this in (10.16), we obtain

$$\xi_i = \frac{i(r-i)}{2p} + \frac{iR}{r}, \tag{10.18}$$

which holds for $0 \leq i \leq r$.

The function $\xi = (\xi_0, \xi_1, \xi_2, \dots)$ given by (10.18), together with $\xi_i = R$ for all $i \geq r$, is clearly bounded. However, we need to check that it satisfies all the optimality equations. It remains to verify the following conditions:

$$\begin{aligned}
\xi_i + \gamma &\leq R + p\xi_1 + q\xi_0 & (i \leq r-1), \\
\xi_i + \gamma &\leq i + p\xi_{i+1} + q\xi_i & (i \geq r).
\end{aligned}$$

Since $\xi_0 = 0$ and $\gamma = p\xi_1$, the first of these inequalities is equivalent to

$$\xi_i \leq R \text{ for } i \leq r-1 \tag{10.19}$$

and it is easy to see that the second is satisfied if $\gamma \leq i$ whenever $i \geq r$, which means that

$$\gamma \leq r. \tag{10.20}$$

According to the formula (10.18), the condition (10.19) requires that $ir \leq 2pR$ when $i \leq r-1$. This holds provided that

$$r(r-1) \leq 2pR.$$

Similarly, by using equation (10.17), we find that condition (10.20) requires that

$$r(r+1) \geq 2pR.$$

The critical level r was chosen to satisfy (10.15), so both the last two inequalities hold. Thus, we have a bounded solution of the optimality equations and Proposition 10.3 can be applied. This confirms that the replacement policy determined by r is optimal and that the minimum long-term average cost is γ, given by (10.17).

10.4   OPTIMAL INVENTORY POLICIES

This section is concerned with one of the fundamental results in stochastic inventory theory. The model described below is a Markov decision process with an infinite state space consisting of the positive and negative integers. It represents the behaviour of an inventory which holds items of a single type and responds to a sequence of independent and identically distributed demands. A standard form of decision procedure for ordering new supplies to replenish the stock is known as an (s, S) policy. Such policies are determined by selecting two integers s and S with $s < S$ and by applying the rule:

> order the number of items needed to bring the level of the inventory
> exactly up to S whenever the current state falls below s.

As we shall see, this means that if a state $i \leq s$ occurs just after a demand is satisfied, there is an instantaneous transition to the state S. The symbol S will not be used to denote the state space in this section, since we are using standard terminology from inventory theory, where s and S are the critical levels.

Studies of inventory models in operational research were among the earliest applications of dynamic programming. These investigations showed, under suitable assumptions, that the best policy in the class of stationary (s, S) policies is also optimal within the much larger class of Markov decision procedures. The fundamental papers by Iglehart [Igl63] established that this is true for the criteria of minimizing the total discounted cost, and also when the aim is to minimize the long-term average cost. However, his proofs are long and difficult, so there have been several subsequent attempts to simplify the arguments and extend the results. We shall concentrate on the problem of minimizing average costs and a recent paper by Zheng [Zhe91], which gives a relatively simple proof of the optimality of (s, S) policies. His proof is much shorter, but it is not easy because of the technicalities involved in verifying the solution of the optimality equations. The approach is based on constructing a solution by using properties of the best (s, S) policy. Unfortunately, this solution is an unbounded function, so Proposition 10.3 cannot be applied directly. The aim here is to give an outline of Zheng's argument, omitting many of the technical details, and explain his ingenious use of a modified problem, for which the optimality equations have a bounded solution.

Consider an inventory model in discrete time. The state variable i is an integer representing the number of items at present in stock if $i \geq 0$ or a shortage if $i < 0$. Demands in successive periods are independent and identically distributed random variables and the distribution of a typical demand X is given by

$$P(X = r) = p_r \qquad (r = 0, 1, 2, \dots).$$

A demand when the inventory level is i produces a transition $i \rightarrow i - r$, with probability p_r, and a shortage occurs at the start of the next period if $r > i$. Backlogging is allowed, but all demands are eventually satisfied. Orders for replenishment can be made at the beginning of any period and they are delivered instantaneously. This assumption of a zero lead time for deliveries can be relaxed, but it will help to simplify the notation here. We suppose further that the cost of ordering m items is $K + \lambda m$, whenever $m > 0$. The expected demand $\mu = \sum r p_r$ for a single period is assumed to be finite and the price λ per item ordered generates a contribution of magnitude $\lambda \mu$ to the long-term average cost per unit time. This has no effect on the relative merits of different ordering policies so, for convenience, it will be assumed that $\lambda = 0$. The cost of ordering any positive number of items is given by the constant $K > 0$. In effect, a state i at the beginning of any period can be replaced instantaneously by $i + m$, for any integer $m > 0$, at cost K. There is no immediate cost if i is left unchanged until the next demand occurs.

The model also includes a one period holding and shortage cost function c_i. This is determined by the inventory level i at the start of the period, after possible ordering and delivery of new items. Thus, there are two kinds of transition. The currrent state i can be replaced immediately by any state $j \geq i$ and the cost is K if $j > i$, or zero if $j = i$. Then a holding and shortage cost c_j is incurred for that period and the next demand produces a new state $k = j - r$, with probability p_r, at the start of the next period.

We may imagine that c_j is the holding cost for storing j items over one period if $j > 0$, but c_j is a shortage cost representing the penalty for unsatisfied demand if $j < 0$. An example is given by

$$c_j = \alpha j \ \text{ if } \ j \geq 0, \quad c_j = -\beta j \ \text{ if } \ j < 0,$$

where α and β are positive constants. In fact, the optimality of (s, S) policies holds for a wide range of holding and shortage cost functions. From now on, suppose that

$$c_j \geq 0 \ \text{ always and } \ c_j \to \infty \ \text{ as } \ |j| \to \infty.$$

Suppose further that, for some state s^0,

$$c_j \geq c_{j+1} \ \text{ for } \ j < s^0, \quad c_j \leq c_{j+1} \ \text{ for } \ j \geq s^0. \tag{10.21}$$

For a stationary (s, S) policy, the long-term average cost can be evaluated by using a well-known result from renewal theory. The behaviour of the random sequence of inventory levels under the policy essentially consists of independent repetitions of a pattern of events which begins in state S and ends with the next recurrence of state S after a new order is delivered. Each cycle contains one order, so the costs include K and a sum of holding and shortage costs over the number of periods required to reduce the inventory level to a state below s, which marks the end of the cycle. Let $g(s, S)$ be the long-term average cost per unit time for the policy. This can be evaluated by applying the elementary renewal theorem: see Ross [Ros70], for example. We obtain

$$g(s, S) = \frac{K + L(s, S)}{M(S - s)}, \tag{10.22}$$

where the functions L and M are defined as follows. For any initial state $i > s$, $L(s, i)$ is the total expected holding and shortage cost incurred before the level falls below s. Similarly, $M(r)$ is the expected time for the inventory level to be reduced by at least r items. The values $L(s, i)$, for $i = s + 1, s + 2, \ldots, S$, can be determined by using the relation

$$L(s, i) = c_i + \sum_{j=0}^{i-s-1} p_j L(s, i - j).$$

(10.23)

The corresponding relation satisfied by the function M is

$$M(r) = 1 + \sum_{j=0}^{r-1} p_j M(r - j).$$

(10.24)

The expected length of a cycle, $M(S - s)$, can be evaluated by applying this for $r = 1, 2, \ldots, S - s$. Equation (10.22) expresses the long-term average cost of the policy as the ratio of the total expected cost to the expected length, for a single cycle.

The analysis of (s, S) policies in Zheng's paper [Zhe91] shows that the critical levels can be chosen to minimize the average cost $g(s, S)$ and that the optimal parameters s^* and S^* also have the properties stated below.

Suppose that the holding and shortage cost function satisfies condition (10.21) and that $c_j \to \infty$ as $|j| \to \infty$. Then there exist states s^* and S^* such that

(i) $g^* = g(s^*, S^*) = \min_{s < S} g(s, S)$;
(ii) $s^* < s^o \le S^*$;
(iii) $c_{s^*} \ge g^* > c_{s^*+1}$ and $c_{S^*} \le g^*$.

Zheng makes use of these properties in his proof that the optimal (s, S) policy also minimizes the long-term average cost within the class of all stationary policies.

Let δ be the function defined by

$$\delta_0 = 0 \quad \text{and} \quad \delta_j = 1 \quad \text{for all} \quad j \ne 0.$$

The optimality equations for this problem are

$$\xi_i + \gamma = \inf_{j \ge i} \left\{ K \delta_{j-i} + c_j + \sum_{r=0}^{\infty} p_r \xi_{j-r} \right\},$$

(10.25)

for each integer i. The expression on the right includes the cost of an order if $j > i$ as well as the holding and shortage cost for the next period. Notice that infinitely many possible order sizes are permitted and we have assumed previously that each decision set D_i is finite. However, this extension of the model does not effect the validity of Proposition 10.3. The crucial requirement is that the solution $\xi = (\ldots \xi_{-1}, \xi_0, \xi_1, \ldots)$ is a bounded function.

We are seeking a solution of the equations (10.25) which corresponds to an (s, S) policy. If there is a solution of this form, it must have $\gamma = g(s, S)$ and the function ξ must satisfy the conditions

$$\xi_i + \gamma = K + c_S + \sum_{r=0}^{\infty} p_r \xi_{S-r}, \tag{10.26}$$

for $i \leq s$, and

$$\xi_i + \gamma = c_i + \sum_{r=0}^{\infty} p_r \xi_{i-r}, \tag{10.27}$$

whenever $i > s$. Since $S > s$, the equation for the case $i = S$ shows that (10.26) can be replaced by $\xi_i = K + \xi_S$ and, for convenience, we will arrange that $\xi_S = 0$. All the equations (10.26) and (10.27) can be satisfied by defining

$$\xi_i = K + L(s, i) - \gamma M(i - s), \tag{10.28}$$

for $i > s$, and $\xi_i = K$ when $i \leq s$. This is not difficult to verify by using the relations (10.23) and (10.24). Since $\gamma = g(s, S)$, equation (10.22) shows that $\xi_S = 0$.

The function ξ associated with an (s, S) policy can be used to construct a solution of the optimality equations (10.25) by fixing the critical levels $s = s^*, S = S^*$, and setting $g = g^*$. However, it is not difficult to see, from equations (10.28) and the assumption that $c_i \to \infty$ as $i \to \infty$, that $\xi_i \to \infty$ as $i \to \infty$. A solution of the optimality equations based on an (s, S) policy must be unbounded, so it cannot lead directly to a proof that the (s^*, S^*) policy is optimal. However, we know that this policy attains the long-term average cost $g^* = g(s, S)$. Hence, it only remains to prove that no other policy can achieve an average cost which is less that g^*.

The modified problem
In his paper, Zheng [Zhe91] introduces a modification of the inventory model by relaxing the constraint that orders must be non-negative. Thus, instantaneous transitions $i \to j$ are permitted to any new state j and the cost is K if $j \neq i$, as before. The optimality equations for the modified problem are

$$\xi_i + \gamma = \inf_j \left\{ K\delta_{j-i} + c_j + \sum_{r=0}^{\infty} p_r \xi_{j-r} \right\}, \tag{10.29}$$

and they differ from the previous equations (10.25) only by removing the constraint that $j \geq i$ on the right. Since the effect of the modifications is to enlarge the class of admissible policies, it is clear that the minimum long-term average cost cannot be increased. The proof that the (s^*, S^*) policy is optimal in the original problem can be completed by showing that equations (10.29) have a bounded solution with $\gamma = g^*$. In fact, this solution corresponds to a policy which differs from the (s^*, S^*) policy only for certain states $i > S^*$.

Let ξ be defined by $\xi_i = K$ for $i \leq s^*$ and

$$\xi_i = K + L\left(s^*, i\right) - \gamma M\left(i - s^*\right) \tag{10.30}$$

for $s^* < i \leq S^*$, where $\gamma = g^*$. Then, for $i = S^*+1, S^*+2, \ldots, \xi_i$ is defined recursively by

$$\xi_i = \min\left\{ K, c_i - \gamma + \sum_{r=0}^{\infty} p_r \xi_{i-r} \right\}. \tag{10.31}$$

Zheng's proof that the function determined by (10.30) and (10.31) is a bounded solution of the optimality equations (10.29) relies on the properties of s^*, S^* and g^* quoted earlier in this section. Notice that, for states $i > S^*$ with $\xi_i = K$, it is optimal to reduce the level of the inventory immediately to S^*. However, such actions cannot be repeated because, once a state $j \leq S^*$ has occurred, the (s^*, S^*) policy ensures that the random process never returns to a state $k > S^*$.

The argument is completed by applying Proposition 10.3 to the modified problem. This proves that its minimum long-term average cost is g^*. Hence, the same is true for the original problem and it follows that the (s^*, S^*) policy is optimal.

11

Statistical Decisions

11.1 INTRODUCTION

We have investigated a wide range of optimal decision problems involving many different deterministic and stochastic models. So far, none of these problems has included any unknown statistical parameters. Ignorance about parameters in the law of motion of a dynamic system can lead to serious complications in the formulation of sequential decision problems. The aim in this final chapter is to illustrate how the methods we have developed can sometimes be successful, and to give a brief indication of the importance of making further progress in this field of sequential analysis.

In previous chapters, we have always relied on an assumption that, under any particular decision procedure, the behaviour of the system corresponds to a well-defined Markov process. Models containing unknown parameters are more difficult to formulate in this way. For example, suppose we are dealing with a dynamic system in discrete time which is fully specified, apart from an unknown real parameter ω. For each possible value of ω, we have a Markov decision process. The Markovian character of the system can be preserved by using a Bayesian approach, which means assuming that the unknown ω has a prior probability density $\pi_0(\omega)$ at time 0. In general, decisions which control the behaviour of the system at times $1, 2, \ldots$ will lead to a succession of posterior densities $\pi_1(\omega), \pi_2(\omega) \ldots$, allowing for any extra information about ω collected as time progresses. In effect, the state of the system at time t includes the posterior density function $\pi_t(\omega)$. The result may well be a Markov decision process whose state space is infinite-dimensional. This difficulty can sometimes be avoided by restricting to a special class of prior distributions known as conjugate priors: see DeGroot [DeG70]. However, even in such cases, statistical decision problems can be difficult to solve.

Statistics is concerned with sampling to collect information about the unknown parameters of a probability distribution or random process, and then using the information to make reasonable inferences about the underlying distribution. Usually, the aim is to estimate the values of the unknown parameters or to test certain hypotheses about them. We shall confine our attention to the problem of testing two simple hypotheses. The standard form of the problem involves a sample of independent observations and the sample size is fixed in advance. The solution, based on the Neyman–Pearson Lemma, is covered in most text books on Statistics: for example, see Hoel [Hoe84]. For our purposes, it will be helpful to establish a Bayesian version of the result and this will be done in Section 11.2.

Roughly speaking, sequential analysis is concerned with problems of statistical inference where the sample size is not given in advance. The sequential probability ratio test is obtained by generalizing the idea of testing two simple hypotheses to allow the choice of sample size to depend on the information collected during the experiment. A Bayesian formulation leads to an optimal stopping problem and the solution, described in Section 11.3, is equivalent to a sequential probability ratio test. The optimality of such tests is a fundamental result in sequential analysis, and it is remarkable that the proof of optimality given by Arrow, Blackwell and Girshick [ABG49] was based on backwards induction. With the advantage of hindsight, it is now easy to recognise the ideas of dynamic programming in their arguments. These methods have since been applied to many other problems in sequential analysis, particularly those arising in medical statistics. For example, consider two alternative medical treatments designed for patients suffering from a certain disease and suppose that their probabilities of success are unknown. The need to find a procedure for allocating one of these treatments to each of a sequence of patients raises both mathematical and ethical problems which are very difficult to solve. It is by no means obvious at what stage the evidence accumulated in a clinical trial justifies a decision to stop the trial and decide that one or the other treatment should be preferred for all future patients. A great deal of research has been carried out in this field. This theoretical work is now beginning to produce applications of sequential allocation procedures which are reasonably good, even if they are not optimal.

11.2   TESTING STATISTICAL HYPOTHESES

Suppose that we are interested in a random variable X and there are two alternative hypotheses about its probability distribution. For convenience, let us assume that X has a probability density function given by either g or h, where g and h are known functions. The probability of any event $[X \in A]$ is determined by

$$\int_A g(x)\,dx \qquad \text{or} \qquad \int_A h(x)\,dx$$

and we assume that the two alternative hypotheses are distinct, in the sense that

$$\int_A g(x)\,dx \neq \int_A h(x)\,dx \tag{11.1}$$

for some A.

Let $x_1, x_2, \ldots, x_t$ be a random sample of independent observations on X and consider the problem of reaching a decision, based on these data, whether the true distribution of X corresponds to g or h. For the present, we suppose that the sample size t is fixed. According to the Neyman–Pearson Lemma, the decision should be based on the product

$$\lambda_t = \prod_{i=1}^{t} \left\{ \frac{h(x_i)}{g(x_i)} \right\}, \tag{11.2}$$

which is known as the probability ratio or likelihood ratio. It is intuitively clear that large values of λ_t favour the hypothesis associated with h and small values support the alternative that g is the true density of the observations. There is a family of optimal tests, each of which is determined by a critical level k and the rule:

decide in favour of the probability density function g or h according as

$$\lambda_t < k \qquad or \qquad \lambda_t > k.$$

The choice of k can be made in several different ways. Instead of reproducing the original form of the fundamental Lemma, it will be useful in what follows to determine the critical level k by introducing prior probabilities for the two hypotheses and costs for reaching an incorrect final decision.

Let a and b be positive constants. Suppose that a is the cost of wrongly rejecting the probability density g and b is the cost of rejecting h when this is the true density underlying the data. We further suppose that our belief in g or h can be represented by probabilities $(1 - \pi)$ and π, respectively. *A priori*, the probability associated with h is π_0 and the prior probability is replaced by the posterior probability π_t, given the observations $x_1, x_2, \ldots, x_t$. This is done by applying Bayes' Theorem, which amounts to conditioning on the data.

First consider the effect of a single observation $X = x$, say. The joint probability element associated with h and the value x is proportional to $\pi_0 h(x)$. The corresponding probability element for the density g and x is proportional to $(1 - \pi_0) g(x)$. It follows that the ratio of the posterior probabilities $\pi = \pi(x)$ and $(1 - \pi)$ is given by

$$\frac{\pi}{(1 - \pi)} = \frac{\pi_0}{(1 - \pi_0)} \frac{h(x)}{g(x)}. \tag{11.3}$$

Hence

$$\pi(x) = \frac{\pi_0 h(x)}{(1 - \pi_0) g(x) + \pi_0 h(x)}. \tag{11.4}$$

It is easy to extend the formulae and obtain the posterior probability π_t obtained from any random sample $x_1, x_2, \ldots, x_t$. After the first observation, we have

$$\frac{\pi_1}{(1 - \pi_1)} = \frac{\pi_0}{(1 - \pi_0)} \frac{h(x_1)}{g(x_1)},$$

and then, after observing x_2,

$$\frac{\pi_2}{(1 - \pi_2)} = \frac{\pi_1}{(1 - \pi_1)} \frac{h(x_2)}{g(x_2)} = \frac{\pi_0}{(1 - \pi_0)} \frac{h(x_1) h(x_2)}{g(x_1) g(x_2)}.$$

Finally, we obtain

$$\frac{\pi_t}{(1 - \pi_t)} = \frac{\pi_0}{(1 - \pi_0)} \lambda_t \tag{11.5}$$

and $\pi_t = \pi_t(x_1, x_2, \ldots, x_t)$ is given by

$$\pi_t = \frac{\pi_0 \lambda_t}{1 - \pi_0 + \pi_0 \lambda_t}. \tag{11.6}$$

One of the advantages of Bayesian methods in statistical inference is that decision procedures which are optimal with respect to the prior distribution can be constructed by comparing the posterior expected losses for different choices of the final decision. In our case, once we have obtained the posterior probability π_t, it is easy to evaluate the expected costs of a decision in favour of g or h. If this choice is g, there is a posterior probability π_t that the decision is incorrect and the expected cost is $b\pi_t$. If h is chosen, the cost is a when it is the wrong decision and the posterior expected cost is $a(1 - \pi_t)$. The optimal decision must correspond to

$$\min \{a(1 - \pi_t), b\pi_t\}.$$

Hence, it is preferable to choose g if $b\pi_t < a(1 - \pi_t)$ and to choose h if $b\pi_t > a(1 - \pi_t)$. In other words, g or h should be chosen, according as

$$\pi_t < \frac{a}{a + b} \qquad \text{or} \qquad \pi_t > \frac{a}{a + b}. \tag{11.7}$$

Either decision is acceptable when equality occurs. It is easily verified by using the formula (11.6) that this decision rule is equivalent to one based on the ratio λ_t, using the critical level

$$k = \frac{a(1 - \pi_0)}{b\pi_0}.$$

The results so far are restricted to the case when the sample size t is fixed in advance. We have shown that the Bayes procedure for discriminating between the hypotheses that the underlying density is either g or h can be expressed in terms of the posterior probability π_t that h is the true density. The critical value of π_t is

$$d = \frac{a}{a + b}$$

and the rule is:

> choose g or h according as
>
> $$0 \leq \pi_t < d \qquad or \qquad d < \pi_t \leq 1.$$

The choice is immaterial if $\pi_t = d$.

Sequential sampling will be investigated in the next section. The problem will be generalized by introducing a sampling cost c per observation and by considering the optimal choice of the sample size, as well as the final decision. As we shall see, this leads to a natural extension of the above decision procedure. There are two critical levels d' and d'' of the posterior probability, such that

$$0 < d' < d < d'' < 1,$$

and the optimal sequential procedure is defined by the following algorithm.

> Let π_t be the posterior probability associated with h after observing $x_1, x_2, \ldots, x_t$. If $\pi_t \leq d'$, terminate the sample and decide in favour of g; if $\pi_t \geq d''$, stop and decide in favour of h. Take another observation x_{t+1} and repeat, using π_{t+1}, if
>
> $$d' < \pi_t < d''. \qquad (11.8)$$

This procedure is equivalent to a sequential probability ratio test based on the sampling rule:

> given a sample of size t, take another observation if and only if
>
> $$k' < \lambda_t < k''.$$

The positive constants k' and k'' are related to π_0, d' and d'' by

$$k' = \frac{(1 - \pi_0)}{\pi_0} \frac{d'}{(1 - d')}, \qquad k'' = \frac{(1 - \pi_0)}{\pi_0} \frac{d''}{(1 - d'')}.$$

Thus, the sample size of the sequential test is a random variable T defined by

$$T = \min \{t \geq 1 : \lambda_t \leq k' \quad or \quad \lambda_t \geq k''\}$$

and the terminal decision is in favour of g or h according as $\lambda_T \leq k'$ or $\lambda_T \geq k''$, respectively.

The error probabilities of the test are defined by

$$\alpha = P\left(\lambda_T \geq k'' \,|g\right), \qquad \beta = P\left(\lambda_T \leq k' \,|h\right).$$

For example, α is the probability of wrongly choosing h when the true density is g. The corresponding expected sample sizes are $E\{T\,|g\}$ and $E\{T\,|h\}$ under the two alternative hypotheses. This statistical test and many other ideas in sequential analysis first appeared in Wald's book [Wal47]. It has optimality properties which were established in a paper by Wald and Wolfowitz [WW48]. Let α^* and β^* be the two error probabilities for any other test, sequential or not, and let T^* be the corresponding sample size. The sequential probability ratio test is optimal in the sense that, among all such tests for which $\alpha^* \leq \alpha$ and $\beta^* \leq \beta$,

$$E\{T^*\,|g\} \geq E\{T\,|g\} \qquad \text{and} \qquad E\{T^*\,|h\} \geq E\{T\,|h\}.$$

In other words, the sequential probability ratio test minimizes the expected sample sizes under both hypotheses. A full proof of this result can be found in the book by Lehmann [Leh59]. Arrow, Blackwell and Girshick [ABG49] gave an alternative proof, using backwards induction and the main part of their argument is described below. It seems appropriate to end this book with what amounts to one of the earliest applications of the method of dynamic programming.

11.3   THE SEQUENTIAL PROBABILITY RATIO TEST

We shall investigate an optimal stopping problem for the Markov process defined by the sequence $\{\pi_t, t \geq 0\}$, for an arbitrary prior probability π_0. The states π_t for $t \geq 1$ are the posterior probabilities associated with the density function h and the state space is the interval $[0, 1]$. We have already introduced costs a and b for incorrect decisions about whether the true underlying probability density is g or h, respectively. For any probability $\pi \in [0, 1]$, the effective cost of reaching a decision without further sampling is given by

$$K\left(\pi\right) = \min\left\{a\left(1 - \pi\right), b\pi\right\}. \tag{11.9}$$

This is the stopping cost function and, for any state π, the optimal terminal decision is g if $\pi < d$ and h if $\pi > d$, where $d = a/\left(a + b\right)$.

Let $c > 0$ be the cost of a single observation and consider the optimal stopping problem specified by the parameters a, b and c. It will be shown that, for any prior probability π_0, the optimal sample size is determined by a sampling rule of the form (11.8). In other words, for any π_0 and any positive cost parameters a, b, c, the Bayes procedure is equivalent to a sequential probability ratio test.

Given any state π, consider the effect of taking a new observation X. This produces a random transition to a new state, according to the formula (11.4). We have $\pi \to \pi^*(X)$, where

$$\pi^*(X) = \frac{\pi h(X)}{(1 - \pi) g(X) + \pi h(X)}. \qquad (11.10)$$

The random variable X, generating this transition, has a distribution which depends on π. Its probability density function is $(1 - \pi) g(x) + \pi h(x)$, since the occurrence of state π corresponds to assigning probabilities $(1 - \pi)$ and π to the possible densities g and h, respectively. The Markov process consists of a sequence of transitions, $\pi_0 \to \pi_1 \to \pi_2$ and so on, determined by the independent observations $x_1, x_2, \ldots$. In general, $\pi_{t+1} = \pi^*(x_{t+1})$ is determined by setting $\pi = \pi_t$ and substituting the new observation $X = x_{t+1}$ in (11.10).

Apart from the fact that the state variable π is continuous, we have an optimal stopping problem similar to those treated in Chapter 7. We are interested in the function

$$f(\pi) = \text{minimum total expected cost,}$$

where the expectation is conditional on a given initial state, $\pi_0 = \pi$, and the total includes both sampling and terminal decision costs. The minimum can be approximated by restricting the sample size to at most n observations and then letting $n \to \infty$. For each $\pi \in [0, 1]$ and any integer $n \geq 0$, let $f_n(\pi)$ be the minimum future expectation, given that the prior probability of h is π and that the sample size must not exceed n. We have $f_0(\pi) = K(\pi)$ and for $n \geq 1$, the principle of optimality shows that

$$f_n(\pi) = \min \left[K(\pi), c + E\{f_{n-1}(\pi^*(X))\} \right]. \qquad (11.11)$$

The expectation is taken with respect to the random variable X. By using (11.10) and the probability density function of X, conditional on π, we obtain

$$E\{f_{n-1}(\pi^*(X))\}$$
$$= \int f_{n-1}\left(\frac{\pi h(x)}{(1 - \pi) g(x) + \pi h(x)}\right) \{(1 - \pi) g(x) + \pi h(x)\} \, dx \qquad (11.12)$$

The integral here is over the range of possible values of X.

It is a straightforward matter to deduce from these definitions that the functions $f_0, f_1, f_2, \ldots$ form a decreasing sequence. We have $f_1(\pi) \leq K(\pi) = f_0(\pi)$. Then

$$E\{f_1(\pi^*(X))\} \leq E\{f_0(\pi^*(X))\}$$

and it follows from the relation (11.11) that $f_2(\pi) \leq f_1(\pi)$, and so on. In general,

$$f_n(\pi) \leq f_{n-1}(\pi) \leq K(\pi)$$

for $n \geq 1$ and all $\pi \in [0, 1]$. All the functions are non-negative, so they have a
limit:

$$f(\pi) = \lim_{n \to \infty} f_n(\pi).$$

Since the convergence is monotone, this limit function is a solution of the
optimality equation

$$f(\pi) = \min[K(\pi), c + E\{f(\pi^*(X))\}]. \tag{11.13}$$

It can be shown that this represents the minimum expected cost which is
attainable by any sampling policy. In order to establish the form of the optimal
sampling procedure, we need a preliminary result which will be used to show
that all the functions $f_n(\pi)$ and their limit function are concave.

Proposition 11.1 *Let $\varphi(\pi)$ be a non-negative concave function for $\pi \in [0, 1]$
and let ψ be the function defined by*

$$\psi(\pi) = E\{\varphi(\pi^*(X))\},$$

where $\pi^*(X)$ is given by (11.10). Then ψ is also a concave function.

Proof According to (11.12),

$$\psi(\pi) = \int \varphi\left(\frac{\pi h(x)}{(1 - \pi)g(x) + \pi h(x)}\right)\{(1 - \pi)g(x) + \pi h(x)\}\, dx.$$

We need to prove that, for any $\pi_1, \pi_2 \in [0, 1]$ and any $p_1, p_2 \geq 0$ with
$p_1 + p_2 = 1$,

$$\psi(p_1\pi_1 + p_2\pi_2) \geq p_1\psi(\pi_1) + p_2\psi(\pi_2). \tag{11.14}$$

Write $\pi = p_1\pi_1 + p_2\pi_2$ and consider

$$\pi^* = \frac{\pi h}{\{(1 - \pi)g + \pi h\}} = \frac{p_1\pi_1 h + p_2\pi_2 h}{\{(1 - \pi)g + \pi h\}}$$

for any fixed value of x, where $g = g(x)$, $h = h(x)$. The last expression for
π^* can be split into two parts in the following way:

$$\pi^* = \frac{p_1\{(1 - \pi_1)g + \pi_1 h\}}{\{(1 - \pi)g + \pi h\}} \frac{\pi_1 h}{\{(1 - \pi_1)g + \pi_1 h\}}$$

$$+ \frac{p_2\{(1 - \pi_2)g + \pi_2 h\}}{\{(1 - \pi)g + \pi h\}} \frac{\pi_2 h}{\{(1 - \pi_2)g + \pi_2 h\}}.$$

Notice that

$$p_1\{(1 - \pi_1)g + \pi_1 h\} + p_2\{(1 - \pi_2)g + \pi_2 h\} = \{(1 - \pi)g + \pi h\}.$$

The above equation for $\psi(\pi)$ can be written as $\psi(\pi) = \int I dx$, where the integrand $I = I(x)$ is

$$I = \{(1 - \pi)g + \pi h\}\,\varphi(\pi^*).$$

Since φ is a concave function, this integrand satisfies

$$I \geq p_1 I_1 + p_2 I_2,$$

where each component I_j is given by

$$I_j = \{(1 - \pi_j)g + \pi_j h\}\,\varphi\left(\frac{\pi_j h}{(1 - \pi_j)g + \pi_j h}\right), \qquad (j = 1, 2).$$

It follows that

$$\int I dx \geq p_1 \int I_1 dx + p_2 \int I_2 dx$$

and, since $\psi(\pi_j) = \int I_j dx$ for each j, this is equivalent to the required inequality (11.14). $\qquad\square$

If the sampling cost c is large, there may be no advantage in sampling. In particular, if $c \geq ab/(a + b)$, then $c \geq K(\pi)$ for $\pi \in [0, 1]$ and it is easy to show that $f(\pi) = K(\pi)$ always holds. In general, it is not easy to find solutions of the optimality equation (11.13). However, we can use the properties of the limit function f to establish the form of the optimal policy. We are now able to show that f is a concave function, which means that the optimal continuation region C is an interval: see Figure 1.

Proposition 11.2 *Consider the sequence of minimum expected cost functions $f_0, f_1, \ldots$ defined by (11.11) and let f be the limit function.*

(i) *The sequence is non-increasing and it converges to a solution of the optimality equation (11.13).*

(ii) *Each function f_n is concave and so is the limit function f.*

(iii) *Define the continuation region C by*
$$C = \{\pi \in [0, 1] : f(\pi) < K(\pi)\}.$$
If C is non-empty, it is an interval:
$$C = (d', d'') \text{ where } 0 < d' < d < d'' < 1.$$

Proof Part (i) has already been proved by using (11.11).

Part (ii) is a consequence of the previous proposition. Clearly $K(\pi)$ is concave and $f_0(\pi) = K(\pi)$. For $n \geq 1$ Proposition 11.1 shows that $E\{f_{n-1}(\pi^*(X))\}$ is a concave function of π whenever $f_{n-1}(\pi)$ is concave.

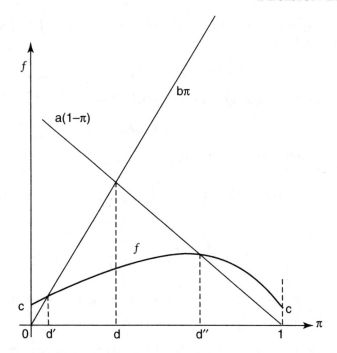

Figure 1. The minimum expected cost function

Then $f_n(\pi)$ is the minimum of two concave functions, so $f_n(\pi)$ is also concave. By induction, each function f_n is concave and it follows that f is a concave function.

For Part (iii), let us assume that $f(\pi) < K(\pi)$ for some $\pi \in (0,1)$ so that C is non-empty. We can use the chord property of concave functions to show that C must be an interval: see Figure 1. For example, we have $0 \leq f(\pi) \leq K(\pi)$ for all π. Since f is concave and $f(0) = f(1) = 0$, it is a continuous function. Further, it is not possible that

$$f(\pi_1) < K(\pi_1), \quad f(\pi_2) = K(\pi_2) \text{ and } f(\pi_3) < K(\pi_3)$$

when $\pi_1 < \pi_2 < \pi_3$. Similarly, the boundary points of C must satisfy $d' < d < d''$. The optimality equation shows that $f(\pi) = K(\pi)$ whenever $K(\pi) < c$, so $d' > 0$ and $d'' < 1$.   □

Suppose that the initial state $\pi_0 \in C$ and that observations $x_1, x_2, \cdots$ lead to posterior probabilities $\pi_1, \pi_2, \ldots$. At each stage, if $\pi_t \in C$, we have $f(\pi_t) < K(\pi_t)$ and there is a definite advantage in observing x_{t+1}. Thus, sampling should continue for an long as $\pi_t \in C$. The optimal sampling procedure is determined by applying this rule and then reaching the appropriate terminal decision. Note that $f(\pi) = b\pi$ for $0 \leq \pi \leq d'$, since $d' < d = a/(a+b)$. Similarly, $f(\pi) = a(1-\pi)$ for $d'' \leq \pi \leq 1$. Sampling continues until the first time that $\pi_t \leq d'$ or $\pi_t \geq d''$ occurs, with a terminal decision in favour of g or h, respectively. This is the Bayes procedure and, as we noted earlier, it is equivalent to a sequential probability ratio test.

Notes on the Exercises

Chapters 1 to 8 each end with a list of Exercises. These notes provide hints towards the solutions and answers, where appropriate.

Chapter 1

1.1 Use induction, or the fact that $\frac{1}{n(n+1)} = \frac{1}{n} - \frac{1}{n+1}$.

1.2 Equation (1.1) shows that

$$
1 + \binom{n+1}{1} + \binom{n+1}{2} + \cdots + \binom{n+1}{n} + 1
$$

$$
= 1 + \left\{ 1 + \binom{n}{1} \right\} + \left\{ \binom{n}{1} + \binom{n}{2} \right\} + \cdots + \left\{ \binom{n}{n-1} + 1 \right\} + 1
$$

$$
= 2 \left\{ 1 + \binom{n}{1} + \binom{n}{2} + \cdots + \binom{n}{n-1} + 1 \right\}.
$$

1.3 If $\sqrt{3} = k_1/k_2$, then $k_1 = 3k_3$ for some integer $k_3 \geq 1$. Hence, $\sqrt{3} = k_2/k_3$ and so on, as in Example 1.3.

1.4 There are two optimal paths attaining the minimum 3.

1.5 There are two shortest paths of length 24.

Chapter 2

2.1 For $n \geq 1$, $f_n(x) = \lambda^{2n-2}x^2$. For $n \geq 2$, $d_n(x) = 0$ and $d_1(x) = x$.

2.2 Show that $f_n(x) = (c_n x)^{\frac{1}{2}}$ for some constant $c_n > 0$. Given c_{n-1}, relation (2.14) determines $f_n(x)$ with $c_n = 1 + \lambda c_{n-1}$. Then (2.15) and (2.16) follow easily.

2.3 $a = 0$, if $0 \leq x \leq \lambda^{-1}\log\lambda$; $\quad a = (\lambda x - \log\lambda)/(1+\lambda)$, if $x > \lambda^{-1}\log\lambda$.

2.4 For each $n \geq 2$, equation (2.20) determines

$$a = -\frac{m}{c+mn}x, \qquad y = x + a = \frac{c + m(n-1)}{c + mn}x.$$

The next step is $d_{n-1}(y)$ and this turns out to be the same as a.

2.5 Define $f_0(x) = 2x$. Then $f_1(x) = 2x + T^{-1}$ and it turns out that $f_n(x) = 2x+$ constant, for any $n \geq 1$. Then we have

$$f_{n+1}(x) = \max\left\{-a^2(T-n) + 2(x+a)\right\} + \text{constant}.$$

This maximum is attained when $a = (T-n)^{-1}$. For $n = T-1$, $f_T(x)$ is attained by choosing $a = 1$. Starting at x_0, the optimal steps are $1, 2^{-1}, 3^{-1}, \dots, T^{-1}$, leading to a final position $x_0 + 1 + 2^{-1} + \cdots + T^{-1}$.

2.6 For fixed n, the minimum cost of reaching 0 from x is $f_n(x) = n + x^2/n$. Now fix x and minimize this by choosing n so that $f_1(x) > f_2(x) > \cdots > f_n(x) \leq f_{n+1}(x)$.

2.7

$$f_1(x) = \min\left\{c|a| + m(x+a)^2\right\}.$$

Let $x > 0$, so that $a \leq 0$ for a minimum. Differentiation shows that the minimum occurs when $x + a = c/2m$ if $x > c/2m$, otherwise $a = 0$. Hence the two formulae for $f_1(x)$. Note that f_1 and its derivative f_1' are continuous at $x = \pm c/2m$.

2.8 This is more difficult. We have

$$f_2(x) = \min\left\{c|a| + f_1(x+a)\right\}$$

and the minimum must occur in the range $-x \leq a \leq 0$, if $x > 0$. The expression on the right is constant if $x + a \geq c/2m$, since $f_1'(x+a) = c$, so we may restrict to choosing some a with $x + a \leq c/2m$. In effect, $f_1(x+a) = f_0(x+a)$ always holds and, as before, the minimum occurs when $x + a = c/2m$ if $x > c/2m$, or when $a = 0$, otherwise. Hence $f_2(x) = f_1(x)$ for all x. Similarly, $f_n(x) = f_1(x)$. The optimal control policy involves at most one step with $a \neq 0$.

Chapter 3

3.1 Shortest path is $1 \to 3 \to 4 \to 5 \to 6 \to 8 \to 9$; length 17.

3.2 Critical path is $1 \to 2 \to 3 \to 5 \to 6 \to 7 \to 9$; length 30.

3.3 Critical path is $1 \to 2 \to 4 \to 5 \to 7$; length 22.

Shortest path is $1 \to 3 \to 5 \to 6 \to 7$; length 11.

3.4 Show successively that $h_{n-1}, h_{n-2}, \dots, h_1$ are uniquely determined.

3.5 Prove that $f_j \geq g_j$ for $j = 1, 2, \dots, n$, by forwards induction. Then show that $f_j \leq h_j$ for $j = n, n-1, \dots, 1$ by backwards induction.

Chapter 4

4.1 The optimal choice is $r_1 = 2, r_2 = 0, r_3 = 1$, giving

$$\sum r_i w_i = 5, \qquad \sum r_i v_i = 160.$$

4.2 The maximum profit is 365. It can be achieved in 3 ways: e.g. send 3 to area 1, 4 to area 2 and 3 to area 3.

4.3

$$f_{n+3}(x) = f_n(x) + 5 \text{ for } n \geq 3.$$

The minimum average cost is $5/3$ and an optimal policy is: replace whenever $x \geq 2$.

4.4 Note that $a_{r-1} \geq a_r$ is equivalent to $r(r+1) \leq 2(k+1)$.

4.5 The given ordering leaves a residual time $t_{10} = 11$ on machine B when the last job is completed on A: total completion time is $\sum a_i + t_{10} = 165$. The optimal ordering reduces t_{10} to 3: minimum total completion time is 157.

4.6 Interchanging machines A and B gives a total completion time of 157, which does not change after applying the algorithm.

4.7 The argument at the end of the proof of Proposition 4.1 shows that Johnson's algorithm produces a schedule in which (4.10) holds whenever i precedes j. Note that (4.10) is equivalent to $\max(-a_i, -b_j) \geq \max(-a_j, -b_i)$ and, by adding $b_i + b_j$, we obtain $t_{ij} \leq t_{ji}$.

Chapter 5

5.1 Use Proposition 5.2 and induction.

5.2 A hint is already given in the exercise.

5.3 Suppose that $h''(x) \geq 0$ always. Then $h'(x)$ exists and it is non-decreasing. Let $x_1 < x_2$, $x_0 = p_1 x_1 + p_2 x_2$, $p_1, p_2 > 0$, $p_1 + p_2 = 1$. There are points x_1' and x_2' such that $x_1 < x_1' < x_0 < x_2' < x_2$ and $h(x_0) - h(x_1) = (x_0 - x_1) h'(x_1')$, $h(x_2) - h(x_0) = (x_2 - x_0) h'(x_2')$. Since $x_0 - x_1 = p_2(x_2 - x_1)$, $x_2 - x_0 = p_1(x_2 - x_1)$ and $h'(x_1') \leq h'(x_2')$, the required inequality (5.1) can be obtained.

5.4 Let $r_j = 2Aj - j^2$ and maximize r_j by using the fact that $r_j \geq r_{j-1}$ if and only if $j \leq A + \frac{1}{2}$.

5.5 Define $f_n(x) = \max \sum_1^n c_j g(a_j)$, subject to $\sum_1^n c_j a_j = x$. Standard techniques lead to

$$f_n(x) = \left(\sum_1^n c_j \right) g \left(x / \sum_1^n c_j \right).$$

Jensen's inequality follows from the fact that, for any $a_1, a_2, \ldots, a_n$, $\sum c_j g(a_j) \leq f_n(\sum c_j a_j)$.

5.6 Let r_k be the reward for allocating the whole quantity x to customer k. If $x > 2c$, then $r_1 > 0$, and the optimal policy is to choose $k = \max\{j \leq n : j(j+1) \leq x/c\}$.

5.7 This is more difficult because the formula for c_n is complicated and best avoided.

    (i) $f_n(x) = \max \{\log a + \theta f_{n-1}(\lambda(x - a))\}$

    (ii)   Standard methods lead to
$$f_n(x) = b_n \log x + c_n, \quad a = d_n(x) = x/b_n.$$

    (iii)   From (ii), $b_n = 1 + \theta b_{n-1}$, so $b_n = (1 - \theta^n)/(1 - \theta)$.

Chapter 6

6.1 We have $g_n(x) = ca^2 + E\{g_{n-1}(x + a + W)\}$ with $a = -\mu$. Then

$$g_0(x) = mx^2, \quad g_1(x) = c\mu^2 + mE\left\{(x - \mu + W)^2\right\},$$
$$g_1(x) = mx^2 + c\mu^2 + m\sigma^2.$$

Since $g_1(x) = g_0(x) + \text{constant}$, the formula for $g_n(x)$ follows easily. As $n \to \infty$, $g_n(x)/n \to c\mu^2 + m\sigma^2$ and $f_n(x)/n \to c\mu^2$.

6.2 When $\frac{1}{2} < p < 1$, $a = (p - q)x$ satisfies $0 < a < x$ and it gives the maximum in (6.12). Then $f_1(x) = \log x + g$ and (6.15) follows without difficulty. The expression $p \log p + (1 - p) \log(1 - p)$ has a unique minimum with respect to p when $p = \frac{1}{2}$. Hence $g > 0$, for $\frac{1}{2} < p < 1$.

6.3 Let f and g be solutions of (6.17) and its boundary conditions. For $j = 1, 2, \ldots, s - 1$, choose a so that it is optimal for f. Then, for some $a \neq 0$,

$$f_j = p f_{j+a} + q f_{j-a}, \quad g_j \geq p g_{j+a} + q g_{j-a}$$

and hence

$$g_j - f_j \geq p \left(g_{j+a} - f_{j+a} \right) + q \left(g_{j-a} - f_{j-a} \right).$$

By considering the smallest difference, $g_j - f_j$, show that $g_j \geq f_j$ always holds. Similarly $g_j \leq f_j$, so $f_j = g_j$ always.

6.4 Use (6.19) and $f_j = (f_j - f_{j-1}) + (f_{j-1} - f_{j-2}) + \cdots + (f_1 - f_0)$ to show that $f_j = f_1 \left\{ 1 + (q/p) + \cdots + (q/p)^{j-1} \right\}$. Then use $f_s = 1$ to determine f_1.

6.5 The strategy $a = \min(j, 5 - j)$, $j = 1, 2, 3, 4$, yields probabilities of success

$$f_1 = p^3 (1 + q) \left(1 - p^2 q^2 \right)^{-1}, \quad f_2 = p^2 (1 + q) \left(1 - p^2 q^2 \right)^{-1},$$

$$f_3 = p \left(1 + p^2 q \right) \left(1 - p^2 q^2 \right)^{-1} \quad f_4 = p (1 + q) \left(1 - p^2 q^2 \right)^{-1}.$$

They satisfy the two optimality conditions if $p < \frac{1}{2}$.

6.6 When $s = 6$, the bold strategy yields

$$f_1 = p^3 (1 - pq)^{-1}, \quad f_2 = p^2 (1 - pq)^{-1}, \quad f_3 = p,$$

$$f_4 = p (1 - pq)^{-1}, \quad f_5 = p \left(1 + q^2 \right) (1 - pq)^{-1}.$$

The optimality conditions $f_3 \geq p f_4 + q f_2$ and $f_3 \geq p f_5 + q f_1$ hold if $p \leq \frac{1}{2}$. The other conditions $f_2 \geq p f_3 + q f_1$, $f_4 \geq p f_5 + q f_3$ are valid for all p.

Chapter 7

7.1 For $n = 1$, $u_i(1) = \max \{ r_i, -c_i + \sum p_{ij} r_j \} \geq u_i(0)$. For $n = 2$, (7.5) and the fact that $u_j(1) \geq u_j(0)$ for all j mean that $u_i(2) \geq u_i(1)$. Similarly, $u_i(3) \geq u_i(2)$ and so on.

7.2 Define $v_i(1) = u_i(1)$ for all i. For $n \geq 2$, set $v_i(n) = r_i$ whenever $i \in S(1)$ and $v_i(n) = -c_i + \sum p_{ij} v_j(n-1)$ if $i \notin S(1)$. Then for $i \notin S(1)$, $v_i(2) = -c_i + \sum p_{ij} u_j(1) \geq -c_i + \sum p_{ij} r_j = u_i(1)$. Hence, $v_i(2) \geq v_i(1)$ for all i and the argument of the previous exercise shows that $v_i(n+1) \geq v_i(n)$ for $n \geq 1$. As $n \to \infty$, $v_i(n) \to v_i$ and this limit is the expectation for the OSLA rule. Clearly $v_i \geq u_i(1)$.

7.3 The function $\{v_i, i \geq 0\}$ was constructed to satisfy

$$v_i = -c + \frac{1}{2}v_i + \frac{1}{2}v_{i+1} \text{ for } i < m \text{ and } v_i = r_i \text{ for } i \geq m.$$

It is a solution of the optimality equation if $v_i \geq r_i$, for $i < m$ and $v_i \geq -c + \frac{1}{2}v_i + \frac{1}{2}v_{i+1}$ for $i \geq m$. For example, the last condition is equivalent to $r_{i+1} - r_i \leq 2c$ and this holds for $i \geq m$, since $2^i \geq r/4c$.

7.4 The rule is to stop as soon as $i \geq m$ and the stopping set is closed. The value function is $v_i = i$ for $i \geq m$ and

$$v_i = \{m\,(m-1) + i\,(i+1)\}\,/2m, \quad \text{for } 1 \leq i \leq m.$$

7.5 Consider Example 7.4 with $a = 6$ and $c = 1$. The optimal strategy corresponds to $m = 3$ and the maximum expectation is $u_j = 9 + j^2$ for $|j| \leq 3$. Replacing the rewards $r_j = 6\,|j|$ by $6\,|j| - 9$ means that $u_0 = 9$ is replaced by zero, so the game is fair.

7.6 The expected time to reach $\pm m$ from $j = 0$ is m^2. Hence, the expected net reward for the policy is $(c + d)\,m^2 - cm^2 = dm^2$. This can be made arbitrarily large by making m large, so the supremum is infinite.

7.7 Taking $f_i\,(0) = 0$ for all i ensures that $f_i\,(1) \geq f_i\,(0)$ and it follows by induction that $f_i\,(n+1) \geq f_i\,(n)$ always. If we start with $f_i\,(0) = s_i$, then $f_i\,(1) \leq f_i\,(0)$ and this leads to a decreasing sequence of approximations.

7.8 The first part is awkward. The inequality can be verified by treating the following cases separately: $\beta \leq \gamma, \alpha < \gamma < \beta, \gamma \leq \alpha \leq \beta$ and $\gamma < \beta < \alpha$. The second inequality follows by using the optimality equation satisfied by f_i and g_i. The exercise contains a hint for the last part.

7.9 Let $\delta_j = w_{j-1} - w_j$ for $j = 0, 1, \ldots, m - 1$. The equation for w_j is equivalent to $\delta_{j+1} = \delta_j + 2$. By symmetry, $w_{-j} = w_j$ and $\delta_0 = -\delta_1$, so $\delta_1 = 1$. Hence, $\delta_2 = 3, \ldots, \delta_j = 2j - 1$. Then $\delta_{j+1} + \delta_{j+2} + \cdots + \delta_m = w_j - w_m$ and $w_m = 0$. For $j \geq 0$,

$$w_j = 2j + 1 + 2j + 3 + \cdots + 2m - 1 = m^2 - j^2.$$

7.10 (i) Clearly $v_1 = p + (1 - p)\,k$, since $k > 1$. The equation for v_i follows from the principle of optimality.

(ii) This part needs a careful argument. The optimality equation shows that $v_i \leq v_{i-1}$ always. More precisely,

$$\text{either } v_i = v_{i-1} \leq i, \text{ or } v_{i-1} > i \text{ and } v_i < v_{i-1}.$$

If $v_i = v_{i-1}$, then $v_{i-1} \leq i$. Hence, $v_i < i+1$ and $v_{i+1} = v_i$. Define m as the largest integer such that $v_{m-1} \geq m$, taking $v_0 = k$ so that $m \geq 1$. Then it follows that $v_j < j+1$ and $v_{j+1} = v_j$ for all $j \geq m$ and it is optimal to avoid parking at any position $> m$.

(iii) $m = 3$.

Chapter 8

8.1 Equation (8.5) takes the form $L(\xi) = c$, where $L'(z) = -\int_z^\infty f(y)\,dy \geq -1$. Then $L(0) = \mu$ and $L(\xi) = L(0) + \int_0^\xi L'(z)\,dz$ for any $\xi > 0$, so $c \geq \mu - \xi$.

8.2 Use the substitution $y = az + b$ in (8.5) and reduce it to $a \exp\{-(\xi - b)/a\} = c$. Then $\xi = a\log(a/c) + b$ and the critical level is about £1600.

8.3 $\xi = (1 - c)/c$.

8.4 Use integration by parts to simplify equation (8.9).

8.5 Waiting for the next offer Z at cost c produces a net expectation $E\{\max(y, Z)\} - c$ and this exceeds y if and only if $\int_y^\infty (z - y) f(z)\,dz > c$, which is equivalent to the condition $y < \xi$.

8.6 Suppose that $g_{r+1} = \frac{r+1}{n}\left\{\frac{1}{r+1} + \frac{1}{r+2} + \cdots + \frac{1}{n-1}\right\}$ for some $r \leq n - 2$. Then (8.19) shows that

$$g_r = \frac{1}{n} + \frac{r}{r+1}g_{r+1} = \frac{r}{n}\left\{\frac{1}{r} + \frac{n}{r+1}g_{r+1}\right\}$$

and (8.20) follows.

8.7 The inequalities here and (8.21) show that

$$\log(n/s) < 1 < \log((n-1)/(s-2)).$$

It follows that s/n lies between $1/e$ and $1/e + c/n$ where $c = 2 - 1/e > 0$. Hence, $s/n \to 1/e$ as $n \to \infty$.

8.8 Suppose that $v_n \geq (n+1)/(n+3)$ for some $n \geq 1$. This is true for $n = 1$. Then $v_{n+1} = (1 + v_n^2)/2$, so $v_{n+1} \geq \left\{1 + (n+1)^2(n+3)^{-2}\right\}/2$ and, after some calculations, it can be verified that $v_{n+1} \geq (n+2)/(n+4)$.

8.9 Equality holds for $n = 1$, since $u_n = v_n = 1$. For $n \geq 2$,

$$\log n = \int_1^n \frac{1}{t}\,dt > \frac{1}{2} + \frac{1}{3} + \cdots + \frac{1}{n},$$

as in Exercise 8.7. Hence $u_n < 1 + \log n$.

Assume that $v_n \geq \log(a + n)$ for some $n \geq 1$ and note the following facts: $v + e^{-v}$ is an increasing function of v for $v > 0$, and $\log(1 + u) < u$ for $u > 0$. Then we have

$$
\begin{aligned}
v_{n+1} \;=\; & v_n + e^{-v_n} \geq \log(a + n) + \frac{1}{a + n} \\
\geq \; & \log(a + n) + \log\left(1 + \frac{1}{a + n}\right) = \log(a + n + 1).
\end{aligned}
$$

References

[ABG49] Arrow K. J., Blackwell D. and Girshick M. J. (1949) Bayes and minimax solutions of sequential decision problems. *Econometrica* 17: 213–244.

[AHBT82] Abdel-Hamid A. R., Bather J. A. and Trustrum G. B. (1982) The secretary problem with an unknown number of candidates. *J. Appl. Prob.* 19: 619–630.

[Bat73] Bather J. A. (1973) Optimal decision procedures for finite markov chains. Parts I and II. *Adv. Appl. Prob.* 5: 328–339 and 521–540.

[Bel57] Bellman R. (1957) *Dynamic Programming.* Princeton University Press, Princeton, N.J.

[Bla62] Blackwell D. (1962) Discrete dynamic programming. *Ann. Math. Statist.* 33: 719–726.

[Boy68] Boyer C. B. (1968) *A History of Mathematics.* Wiley, New York.

[DeG70] DeGroot M. H. (1970) *Optimal Statistical Decisions.* McGraw-Hill, New York.

[Fel68] Feller W. (1968) *An Introduction to Probability Theory and Its Applications,* volume I. Wiley, New York, third edition.

[Fre82] French S. (1982) *Sequencing and Scheduling.* Ellis Horwood, Chichester.

[Fre83] Freeman F. R. (1983) The secretary problem and its extensions: review. *Int. Statist. Rev.* 51: 189–206.

[GS82] Grimmett G. R. and Stirzaker D. R. (1982) *Probability and Random Processes.* Oxford University Press, Oxford.

[HK92] Hill T. P. and Kertz R. P. (1992) A survey of prophet inequalities in optimal stopping theory. *Contemp. Math.* 125: 191–207.

[Hoe84] Hoel P. G. (1984) *Introduction to Mathematical Statistics.* Wiley, New York.

[How60] Howard R. A. (1960) *Dynamic Programming and Markov Processes.* MIT Press and Wiley, New York.

[Igl63] Iglehart D. (1963) Dynamic programming and stationary analysis of inventory problems. Chapter in *Multistage Inventory Models and Techniques.* Stanford University Press.

[KS60] Kemeny J. G. and Snell J. L. (1960) *Finite Markov Chains.* Van Nostrand, New York.

[Leh59] Lehmann E. L. (1959) *Testing Statistical Hypotheses.* Wiley, New York.

[Put94] Puterman M. L. (1994) *Markov Decision Processes.* Wiley, New York.

[Ros70] Ross S. M. (1970) *Applied Probability Models with Optimization Applications.* Holden–Day, San Francisco.

[Ros83] Ross S. M. (1983) *Introduction to Stochastic Dynamic Programming.* Academic Press, New York.

[RPS87] Ravindran A., Philips D. T. and Solberg J. J. (1987) *Operations Research, Principles and Practice.* Wiley, New York, second edition.

[van84] van Tiel J. (1984) *Convex Analysis: An Introductory Text.* Wiley, Chichester.

[Wal47] Wald A. (1947) *Sequential Analysis.* Wiley, New York.

[Whi82] Whittle P. (1982) *Optimization over Time, Dynamic Programming and Stochastic Control,* volume I. Wiley, New York.

[Whi83] Whittle P. (1983) *Optimization over Time, Dynamic Programming and Stochastic Control,* volume II. Wiley, New York.

[WW48] Wald A. and Wolfowitz J. (1948) Optimum character of the sequential probability ratio test. *Ann. Math. Statist.* 19: 326–339.

[Zhe91] Zheng Y. S. (1991) A simple proof for optimality of (s, S) policies in infinite horizon inventory systems. *J. App. Prob.* 28: 802–810.

Index